Handbook of Poultry Practice

Handbook of Poultry Practice

Giriraj Prajapati

RANDOM

RANDOM PUBLICATIONS
NEW DELHI (INDIA)

Handbook of Poultry Practice

ISBN 978-93-5111-380-5

Published in 2014 in India by

RANDOM PUBLICATIONS

4376-A/4B, Gali Murari Lal, Ansari Road
New Delhi-110 002
Phone : +91-11-43580356, +91-11-23289044
e-mail: randomexports@gmail.com, sales@randompublications.com, info@randompublications.com

Reprinted 2026

Type Setting by: Friends Media, Delhi-110089

Preface

Poultry farming, raising of birds domestically or commercially, primarily for meat and eggs but also for feathers. Chickens, turkeys, ducks, and geese are of primary importance, while guinea fowl and squabs are chiefly of local interest. This article treats the principles and practices of poultry farming. For a discussion of the food value and processing of poultry products. Humans first domesticated chickens of Indian origin for the purpose of cockfighting in Asia, Africa, and Europe. Very little formal attention was given to egg or meat production. Cockfighting was outlawed in England in 1849 and in most other countries thereafter. Exotic breeds and new standard breeds of chickens proliferated in the years to follow, and poultry shows became very popular. From 1890 to 1920 chicken raisers stressed egg and meat production, and commercial hatcheries became important after 1920.

Chicken feeding is a highly perfected science that ensures a maximum intake of energy for growth and fat production. High quality and well-balanced protein sources produce a maximum amount of muscle, organ, skin, and feather growth. The essential minerals produce bones and eggs, 3 to 4 percent of the live bird being composed of minerals and 10 percent of the egg. Calcium, phosphorus, sodium, chlorine, potassium, sulphur, manganese, iron, copper, cobalt, magnesium, and zinc are all required. Vitamins A, C, D, E and K and all 12 of the B vitamins are also required. Water is essential, and antibiotics are almost universally used to stimulate appetite, control harmful bacteria, and prevent disease. Modern rations produce a pound of broiler on about 2 pounds (0.9 kg) of feed and a dozen eggs from 4.5 pounds (2 kg) of feed. Chicken breeding is an outstanding example of the application of basic genetic principles of inbreeding, linebreeding, and crossbreeding, as well as of intensive mass selection to effect faster and cheaper gains in broilers and maximum egg production for the egg-laying strains. Maximum use of heterosis, or hybrid vigour, through incrosses and crossbreeding has been made. Crossbreeding for egg production has used the single-comb White Leghorn, the Rhode Island Red, the New Hampshire, the Barred Plymouth Rock, the White Plymouth Rock, the Black Australorp, and the White Minorca. Crossbreeding for broiler production has used the White Plymouth Rock or New Hampshire crossed with White or Silver Cornish or incrosses utilizing widely diverse inbred strains within a single breed. Rapid and efficient weight gains, and high quality, plump, meaty carcasses have been achieved thereby. The male sperm lives in the hen's oviduct for two to three weeks. Eggs are fertilized within 24 hours after mating. Yolks originate in the ovary and grow to about 1.6 inches (4 cm) in diameter, after which they are released into the oviduct, where the thick white and two shell membranes are added. The egg then moves into the uterus where the thin white and the shell are added. This process requires a total of 24 hours per egg. The hatching of fertilized eggs requires 21 days, with the heavy breeds requiring a few more hours and the lighter breeds slightly fewer. Ideal hatching temperature approximates 100 °F (38 °C) with control of air flow, humidity, oxygen, and carbon dioxide being essential. Standardized egg-laying tests and official random sample tests have been used for many years to measure actual productivity.

The present book is a concise and up-to-date work including all aspects of this subject. Every effort has been made in the representation manner in this book so that the students do not get confused while going through this book.

I thank all members of my team who have helped in the preparation of the book. My special thanks go to "Random Publication" who have published the book.

—Giriraj Prajapati

Contents

1

Poultry Farming Techniques

Poultry farming is the practice of raising domesticated birds such as chickens, turkeys, ducks, and geese, as a subcategory of animal husbandry, for the purpose of farming meat or eggs for food. More than 50 billion chickens are raised annually as a source of food, for both their meat and their eggs.

Chickens raised for meat are called broilers, whilst those raised for eggs are called laying hens. In total, the UK alone consumes over 29 million eggs per day. Some hens can produce over 300 eggs a year. Chickens will naturally live for 6 or more years. After 12 months, the hen's productivity will start to decline. This is when most commercial laying hens are slaughtered.

The majority of poultry are raised using intensive farming techniques. According to the Worldwatch Institute, 74 percent of the world's poultry meat, and 68 percent of eggs are produced this way. One alternative to intensive poultry farming is free range farming.

Friction between these two main methods has led to long term issues of ethical consumerism. Opponents of intensive farming argue that it harms the environment and creates health risks, as well as abusing the animals themselves.

Advocates of intensive farming say that their highly efficient systems save land and food resources due to increased productivity, stating that the animals are looked after in state-of-the-art environmentally controlled facilities. A few countries have banned cage system housing, including Sweden and Switzerland. Consumers can still purchase lower cost eggs from other countries' intensive poultry farms.

Techniques

Free-range

Free-range poultry farming consists of poultry permitted to roam freely instead of being contained in any manner. In the UK, the Department for Environment, Food and Rural Affairs says that a free range chicken must have daytime access to open-air runs during at least half of its life. Unlike in the United States, this definition also applies to eggs. The European Union regulates marketing standards for egg farming which specifies a minimum condition for Free Range Eggs states that "hens have continuous daytime access to open-air runs, except in the case of temporary restrictions imposed by veterinary authorities". In free-range broiler systems, the chickens are given continuous access to an outdoor range during the daytime and sheds where they are housed at night. Free-range chickens grow more slowly than intensive chickens. They live at least 56 days. In the EU, each chicken must have one square metre of outdoor space.

Figure: *Free range chickens being fed outdoors*

Free-range poultry production requires that the poultry have access to the outside. In some cases this means the poultry are raised on pasture, enabling the poultry to move around, forage for their natural diet and live in cleaner conditions than those in batteries. In some farms, the manure from free-range poultry can be used to benefit crops.

The benefits are also an increased growth rate and opportunities for natural behaviour such as pecking, scratching, foraging and exercise outdoors, as well as fresh air and daylight. Because they grow slower and have opportunities for exercise, free-range chickens have better leg and heart health and a much higher quality of life.

Finding suitable land with adequate drainage to minimise worms and coccidial oocysts, suitable protection from prevailing winds, good ventilation, access and protection from predators can be difficult. Excess heat, cold or damp can have a harmful effect on the animals and their productivity. Unlike battery farms, free range farmers have little control over the food their animals come across, which can lead to unreliable productivity.

Some free range farming in the UK, which accounts for 26% of production, has also come under criticism concerning animal welfare. This is due to some large-scale free range farms where social abnormalities arise due to having large numbers of birds in an outdoor space. Beak trimming due to cannibalism and infighting is common in this form of poultry farming as well as in batteries. Diseases are common and the animals are vulnerable to predators. In South-East Asia, a lack of disease control in free range farming has been associated with outbreaks of Avian influenza.

In organic systems, chickens are also free-range. Organic chickens are slower growing, more traditional breeds and live typically for around 81 days. They grow at half the rate of intensive chickens. They have a larger space allowance outside (at least 2 square metres and sometimes up to 10 square metres per bird).

Yarding

While often confused with free-range farming, yarding is actually a separate method of poultry culture by which chickens and cows are raised together. The distinction is that free-range poultry are either totally unfenced, or the fence is so distant that it has little influence on their freedom of movement. Yarding is common technique used by small farms in the Northeastern US.

Daily releases out of hutches or coops allows for instinctual nature for the chickens with protections from predators. The hens usually lay eggs either on the ground of the coop or in baskets if provided by the farmer. This technique can be complicated if used with roosters though, mostly because of difficulty getting them into the coop and to clean the coop while it is inside. This territorial nature is apparent while outside in which they have a brood of hens and sometimes even informal land claims. This can endanger people unaware of the existence of the territories who are attacked by the larger birds.

Figure: *Ducks and other poultry*

Intensive Chicken Farming

In egg-producing farms, birds are typically housed in rows of battery cages. Environmental conditions are automatically controlled, including light duration, which mimics summer daylength. This stimulates the birds to continue to lay eggs all year round. Normally, significant egg production only occurs in the warmer months. Critics argue that year-round egg production stresses the birds more than normal seasonal production.

Figure: *Egg-laying chickens in battery cages*

Meat chickens, commonly called broilers, are floor-raised on litter such as wood shavings or rice hulls, indoors in climate-controlled housing.

Poultry producers routinely use nationally approved medications, such as antibiotics, in feed or drinking water, to treat disease or to prevent disease outbreaks arising from overcrowded or unsanitary conditions. In the U.S., the national organisation overseeing chicken production is the Food and Drug Administration (F.D.A.). Some F.D.A.-approved medications are also approved for improved feed utilisation.

Figure: *Egg-laying chicken 5 days out of battery cage*

Figure: *Broilers in a production house*

In egg-producing farms, cages allow for more birds per unit area, and this allows for greater productivity and lower space and food costs, with more efforts put into egg-laying.

In the U.S., for example, the current recommendation by the United Egg Producers is 67 to 86 in^2 (430 to 560 cm^2) per bird, which is about 9 inches by 9 inches. Modern poultry farming is very efficient and allows meat and eggs to be available to the consumer in all seasons at a lower cost than free range production, and the poultry have no exposure to predators.

The cage environment of egg producing does not permit birds to roam. The closeness of chickens to one another frequently causes cannibalism. Cannibalism is controlled by de-beaking (removing a portion of the bird's beak with a hot blade so the bird cannot effectively peck). However, de-beaking does not fully prevent cannibalism it just reduces the damage. Most battery chickens are missing 30-70% of plumage by the time that they are spent.

Another condition that can occur in prolific egg laying breeds is osteoporosis. This is caused from year-round rather than seasonal egg production, and results in chickens whose legs cannot support them and so can no longer walk. During egg production, large amounts of calcium are transferred from bones to create eggshell. Although dietary calcium levels are adequate, absorption of dietary calcium is not always sufficient, given the intensity of production, to fully replenish bone calcium.

Under intensive farming methods, a meat chicken will live less than six weeks before slaughter. This is half the time it would take traditionally. This compares with free-range chickens which will usually be slaughtered at 8 weeks, and organic ones at around 12 weeks.

In intensive broiler sheds, the air can become highly polluted with ammonia from the droppings.

This can damage the chickens' eyes and respiratory systems and can cause painful burns on their legs (called hock burns) and feet. Chickens bred for fast growth have a high rate of leg deformities because they cannot support their increased body weight. Because they cannot move easily, the chickens are not able to adjust their environment to avoid heat, cold or dirt as they would in natural conditions. The added weight and overcrowding also puts a strain on their hearts and lungs. In the U.K., up to 19 million chickens die in their sheds from heart failure each year.

Indoor with Higher Welfare

Chickens are kept indoors but with more space (around 12 to 14 birds per square metre). They have a richer environment for example with natural light or straw bales that encourage foraging and perching.

The chickens grow more slowly and live for up to two weeks longer than intensively farmed birds. The benefits of higher welfare indoor systems are the reduced growth rate, less crowding and more opportunities for natural behaviour.

Issues with Poultry Farming

Humane Treatment

Animal welfare groups have frequently criticized the poultry industry for engaging in practices which they believe to be inhumane. Many animal rights advocates object to killing chickens for food, the "factory farm conditions" under which they are raised, methods of transport, and slaughter. Compassion Over Killing and other groups have repeatedly conducted undercover investigations at chicken farms and slaughterhouses which they allege confirm their claims of cruelty. Conditions in intensive chicken farms may be unsanitary, allowing the proliferation of diseases such as salmonella and E. coli.

Chickens may be raised in total darkness; hens are most often kept in crowded wire battery cages with space less than that of a sheet of paper per hen, as opposed to cage-free or free range.

Rough handling and crowded transport during various weather conditions and the failure of existing stunning systems to render the birds unconscious before slaughter have also been cited as welfare concerns.

Another animal welfare concern is the use of selective breeding to create heavy, large-breasted birds, which can lead to crippling leg disorders and heart failure for some of the birds. Concerns have been raised that companies growing single varieties of birds for eggs or meat are increasing their susceptibility to disease.

A common practice among hatcheries is the culling of newly born male chicks of egg laying breeds, since they don't lay eggs, and do not grow fast enough to be profitable for meat.

Debeaking

Laying hens are routinely de-beaked when young to prevent fighting and feather pecking. Animal rights activist claim this is bad because beaks are sensitive, and the usual practice of trimming them without anaesthesia is considered inhumane by some.

De-beaked chickens will peck much less than chickens with beaks, which animal behaviourist Temple Grandin attributes to guarding against pain. The chicken industry says that de-beaking is not painful.

Others argue that the procedure causes lifelong chronic pain and discomfort and decreased ability to eat or drink.

Intelligence

Some groups which advocate for more humane treatment of chickens claim that chickens are intelligent. Dr. Chris Evans of Macquarie University claims that their range of 20 calls, problem solving skills, use of representational signalling, and the ability to recognise each other by facial features demonstrate the intelligence of chickens.

Antibiotics

Antibiotics have been used on poultry in large quantities since the 1940s, when it was found that the by products of antibiotic production, fed because the antibiotic-producing mould had a high level of vitamin B_{12} after the antibiotics were removed, produced higher growth than could be accounted for by the vitamin B_{12} alone. Eventually it was discovered that the trace amounts of antibiotics remaining in the by products accounted for this growth.

The mechanism is apparently the adjustment of intestinal flora, favouring "good" bacteria while suppressing "bad" bacteria, and thus the goal of antibiotics as a growth promoter is the same as for probiotics. Because the antibiotics used are not absorbed by the gut, they do not put antibiotics into the meat or eggs.

Antibiotics are used routinely in poultry for this reason, and also to prevent and treat disease.

Many contend that this puts humans at risk as bacterial strains develop stronger and stronger resistances. Critics point out that, after six decades of heavy agricultural use of antibiotics, opponents of antibiotics must still make arguments about theoretical risks, since actual examples are hard to come by. Those antibiotic-resistant strains of human diseases whose origin is known originated in hospitals rather than farms.

A proposed bill in the United States Congress would make the use of antibiotics in animal feed legal only for therapeutic (rather than preventative) use, but it has not been passed. However, this may present the risk of slaughtered chickens harbouring pathogenic bacteria and passing them on to humans that consume them.

In October 2000, the U.S. Food and Drug Administration (FDA) discovered that two antibiotics were no longer effective in treating diseases found in factory-farmed chickens; one antibiotic was swiftly pulled from the market, but the other, Baytril, was not. Bayer, the company which produced it, contested the claim and as a result, Baytril remained in use until July 2005.

To prevent any residues of antibiotics in chicken meat, any given antibiotics are required to have a "withdrawal" period before they can be slaughtered. Samples of poultry at slaughter are randomly tested by the FSIS, and shows a very low percentage of residue violations

Arsenic

Chicken feed can also include Roxarsone, an antimicrobial drug that also promotes growth. Roxarsone was used as a broiler starter by about 70% of the broiler growers between 1995 to 2000. The drug has generated controversy because it contains arsenic, which is highly toxic to humans. This arsenic could be transmitted through runoff from the poultry yards. A 2004 study by the U.S. magazine Consumer Reports reported "no detectable arsenic in our samples of muscle" but found "A few of our chicken-liver samples has an amount that according to EPA standards could cause neurological problems in a child who ate 2 ounces of cooked liver per week or in an adult who ate 5.5 ounces per week." The U.S. Food and Drug Administration (FDA), however, is the organisation responsible for the regulation of foods in America, and all samples tested were "far less than the amount allowed in a food product."

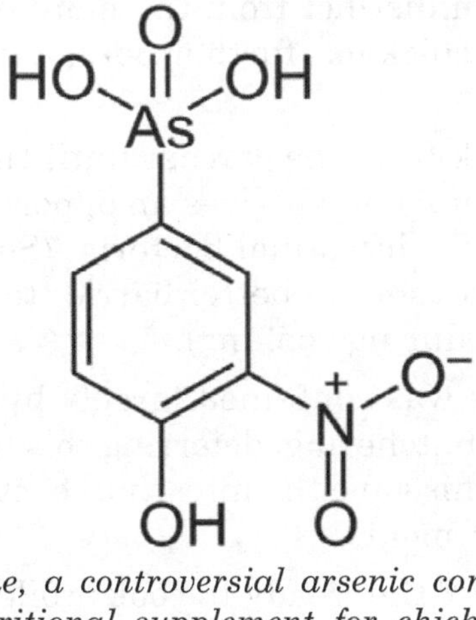

Figure: *Roxarsone, a controversial arsenic compound used as a nutritional supplement for chickens.*

Growth Hormones

Hormone use in poultry production is illegal in the United States. Similarly, no chicken meat for sale in Australia is fed hormones.

Several scientific studies have documented the fact that chickens grow rapidly because they are bred to do so, not because of growth hormones. A small producer of natural and organic chickens confirmed this assumption:

"If this were 1948, you might have something to worry about. Using hormones to boost egg production was a brief fad in the Forties, but was abandoned because it didn't work. Using hormones to produce soft-meated roasters was used to some extent in the Forties and Fifties, but the increased growth rates of broilers made the practice irrelevant—the broilers got as big as anyone wanted them to get when they were still young enough to be soft-meated without chemicals.

The only hormone that was ever used in any quantity on poultry (DES) was banned in 1959, after everyone but a few die-hard farmers had given them up as a silly idea. Hormones are now illegal in poultry and eggs. The people who advertise "No hormones" are either woefully ignorant or are indulging in cynical fear-mongering, maybe both."

E. Coli

According to Consumer Reports, "1.1 million or more Americans are sickened each year by undercooked, tainted chicken." A USDA study discovered *E. coli* in 99% of supermarket chicken, the result of chicken butchering not being a sterile process. However, the same study also cautions that the type of *E. coli* turned up was in every case a non-lethal form distinct from the more dangerous "O157:H7" strain. Many of these chickens, furthermore, had relatively low levels of contamination.

Feces tend to leak from the carcass until the evisceration stage, and the evisceration stage itself gives an opportunity for the interior of the carcass to receive intestinal bacteria. (So does the skin of the carcass, but the skin presents a better barrier to bacteria and reaches higher temperatures during cooking).

Before 1950, this was contained largely by not eviscerating the carcass at the time of butchering, deferring this until the time of retail sale or in the home. This gave the intestinal bacteria less opportunity to colonise the edible meat.

The development of the "ready-to-cook broiler" in the 1950s added convenience while introducing risk, under the assumption that end-to-end refrigeration and thorough cooking would provide adequate protection. *E. coli* can be killed by proper cooking times, but there is still some risk associated with it, and its near-ubiquity in commercially farmed chicken is troubling to some. Irradiation has been proposed as a means of sterilising chicken meat after butchering.

Avian Influenza

There is also a risk that crowded conditions in chicken farms will allow avian influenza (bird flu) to spread quickly. A United Nations press release states: "Governments, local authorities and international agencies need to take a greatly increased role in combating the role of factory-farming, commerce in live poultry, and wildlife markets which provide ideal conditions for the virus to spread and mutate into a more dangerous form..."

Efficiency

Farming of chickens on an industrial scale relies largely on high protein feeds derived from soybeans; in the European Union the soybean dominates the protein supply for animal feed, and the poultry industry is the largest consumer of such feed. Two kilograms of grain must be fed to poultry to produce 1 kg of weight gain. However, for every gram of protein consumed, chickens yield only 0.33 g of edible protein.

Economic Factors

Changes in commodity prices for poultry feed have a direct effect on the cost of doing business in the poultry industry. For instance, a significant rise in the price of corn in the United States can put significant economic pressure on large industrial chicken farming operations.

World Chicken Population

The Food and Agriculture Organisation of the United Nations estimated that in 2002 there were nearly sixteen billion chickens in the world, counting a total population of 15,853,900,000. The figures from the *Global Livestock Production and Health Atlas* for 2004 were as follows:

1. China (3,860,000,000)
2. United States (1,970,000,000)
3. Indonesia (1,200,000,000)
4. Brazil (1,100,000,000)
5. Mexico (540,000,000)
6. India (495,000,000)
7. Russia (340,000,000)

8. Japan (286,000,000)
9. Iran (280,000,000)
10. Turkey (250,000,000)
11. Bangladesh (172,630,000)
12. Nigeria (143,500,000)

Poultry

Poultry is a category of domesticated birds kept by humans for the purpose of collecting their eggs, or raising for their meat and/or feathers. These most typically are members of the superorder Galloanserae (fowl), especially the order Galliformes (which includes chickens, quails and turkeys) and the family Anatidae (in order Anseriformes), commonly known as "waterfowl" (e.g. domestic ducks and domestic geese).

Poultry also includes other birds which are killed for their meat, such as pigeons or doves or birds considered to be game, like pheasants. Poultry comes from the French/Norman word, poule, itself derived from the Latin word Pullus, which means small animal.

Poultry is the second most widely eaten meat in the world, accounting for about 30% of meat production worldwide, after pork at 38%.

Cuts of Poultry

The meatiest parts of a bird are the flight muscles on its chest, called breast meat, and the walking muscles on the first and second segments of its legs, called the thigh and drumstick, respectively. The wings are also eaten, usually (in the United States) without separating them, as in Buffalo wings; the first and second segment of the wings are referred to as drumette (meatier) and flat when these need to be distinguished, though these are technical terms.

In Japan, the wing is frequently separated, and these parts are referred to as Kb½nCQ (*teba-moto* "wing base") and Kb½nHQ (*teba-saki* "wing tip"). Dark meat, which avian myologists refer to as "red muscle," is used for sustained activity—chiefly walking, in the case of a chicken.

The dark colour comes from the protein myoglobin, which plays a key role in oxygen uptake within cells. White muscle, in contrast, is suitable only for short, ineffectual bursts of activity such as, for chickens, flying. Thus the chicken's leg and thigh meat are dark while its breast meat (which makes up the primary flight muscles) is white. Other birds with breast muscle more suitable for sustained flight, such as ducks and geese, have red muscle (and therefore dark meat) throughout.

Figure: *Cuts from a plucked chicken*

Figure: *The Poultry-dealer, after Cesare Vecellio*

Health

Consumption of large quantities of meat, including poultry, like overconsumption of any caloric food, has certain adverse effects which can include: obesity, heart disease, and constipation. In recent years, health concerns have been raised about the consumption of meat increasing the risk of cancer. Bird and animal fat, particularly from ruminants, tends to have a higher percentage of saturated fat vs. monounsaturated and polyunsaturated fat when compared to vegetable fats, with the exception of some tropical plant fats; consumption of which has been correlated with various health problems. The saturated fat found in meat has been associated with significantly raised risks of colon cancer, although evidence suggests that risks of prostate cancer are unrelated to animal fat consumption. USDA claims that consumption of meat as a source of protein in the human diet is crucial have been resoundingly contradicted by recent studies. The correlation of meat, including poultry, consumption to increased risk of heart disease is controversial.

A survey conducted in 1960 of 25,153 California Seventh-Day Adventists found that the risk of heart disease is three times greater for 45-64 year old men who eat meat daily, versus those who did not eat meat. In another study in 2010 involving over one million people who ate meat found that only processed meat had an adverse risk in relation to coronary heart disease.

The study suggests that eating 50g (less than 2oz) of processed meat per day increases risk of coronary heart disease by 42%, and diabetes by 19%. Chicken meat contains about two to three times as much polyunsaturated fat than most types of red meat when measured as weight percentage. A recent study by the Translational Genomics Research Institute showed that nearly half (47%) percent of the meat and poultry in U.S. grocery stores were contaminated with *Staphylococcus aureus*, with more than half (52%) of those bacteria resistant to antibiotics.

Chicken (Food) Chicken the Most Common Type of Poultry in the World

The modern chicken is a descendant of Red Junglefowl hybrids along with the Grey Junglefowl first raised thousands of years ago in the northern parts of the Indian subcontinent.

Chicken as a meat has been depicted in Babylonian carvings from around 600 BC. Chicken was one of the most common meats available in the Middle Ages. It was widely believed to be easily digested and considered to be one of the most neutral foodstuff. It was eaten over most of the Eastern hemisphere and a number of different kinds of chicken such as capons, pullets and hens were eaten. It was one of the basic ingredients in the so-called white dish, a stew usually consisting of chicken and fried onions cooked in milk and seasoned with spices and sugar.

Chicken consumption in the US increased during World War II due to a shortage of beef and pork. In Europe, consumption of chicken overtook that of beef and veal in 1996, linked to consumer awareness of Bovine spongiform encephalopathy or B.S.E.

Breeding

Modern varieties of chicken such as the Cornish Cross, are bred specifically for meat production, with an emphasis placed on the ratio of feed to meat produced by the animal. The most common breeds of chicken consumed in the US are Cornish and White Rock. Chickens raised specifically for food are called broilers. In the United States, broilers are typically butchered at a young age. Modern Cornish Cross hybrids, for example, are butchered as early as 8 weeks for fryers and 12 weeks for roasting birds. Capons (castrated cocks) produce more and fattier meat. For this reason, they are considered a delicacy and were particularly popular in the Middle Ages.

Edible Components

Figure: Chicken in a public market

Main:

- Breast: These are white meat and are relatively dry.
- Leg: Also called the "drumstick", this is dark meat and is the lower part of the leg.
- Thigh: Also dark meat, this is the upper part of the leg.
- Wing: Comprises three segments, one, shaped like a small drumstick, the middle segment, containing two bones, and the tip, sometimes discarded. Wings are often served as a light meal or bar food. Buffalo wings are a typical example.

Exotic:

- Chicken feet: These contain relatively little meat, and are eaten mainly for the skin and cartilage. Although considered exotic in Western cuisine, the feet are common fare in other cuisines, especially in the Caribbean and China.
- Head: Considered a delicacy in China, the head is split down the middle, and the brains and other tissue is eaten.
- Neck: This is served in various Asian dishes.
- Oysters: Located on the back, near the thigh, these small, round pieces of dark meat are often considered to be a delicacy.
- Pygostyle (chicken's buttocks) and testicles: These are commonly eaten in East Asia and some parts of South East Asia.

By-products:

- Carcase: After the removal of the flesh, this is used for soup stock.
- Chicken eggs
- Heart and gizzard
- Liver: This is the largest organ of the chicken, and is used in such dishes as Pâté and chopped liver.
- Schmaltz: This is produced by rendering the fat, and is used in various dishes.

Health Issues

Chicken meat contains about two to three times as much polyunsaturated fat than most types of red meat when measured as weight percentage.

Chicken generally includes low fat in the meat itself (castrated roosters excluded), however it is highly concentrated on its skin, which should be avoided when low intake of fat is necessary.

Cooking method is highly related to health issues related to chicken, with steam cooking with skin removed being considered one of the healthiest, and fried with trans fats one of the worst.

However according to a 2006 Harvard School of Public Health study of 135,000 people, people who ate grilled skinless chicken 5 or more times a week had a 52 percent higher chance of developing bladder cancer compared to people who didn't. However, such strong associations were not found in individuals regularly consuming chicken with skin intact. A recent study by the Translational Genomics Research Institute showed that nearly half (47%) percent of the meat and poultry in US grocery stores were contaminated with *S. aureus*, with more than half (52%) of those bacteria resistant to antibiotics.

Marketing and Sales

Juvenile chickens, of less than 28 days of age at slaughter in the United Kingdom are marketed as poussin. Mature chicken is sold as small, medium or large.

Figure: *Oven roasted chicken with potatoes.*

Whole mature chickens are marketed in the United States as fryers, broilers, and roasters. Fryers are the smallest size (2.5-4 lbs dressed for sale), and the most common, as chicken reach this size quickly (about 7 weeks). Most dismembered packaged chicken would be sold whole as fryers. Broilers are larger than fryers.

They are typically sold whole. Roasters, or roasting hens, are the largest chickens commonly sold (3-5 months and 6-8 lbs) and are typically more expensive. Even larger and older chickens are called stewing chickens but these are no longer usually found commercially.

The names reflect the most appropriate cooking method for the surface area to volume ratio. As the size increases, the volume (which determines how much heat must enter the bird for it to be cooked) increases faster than the surface area (which determines how fast heat can enter the bird). For a fast method of cooking, such as frying, a small bird is appropriate: frying a large piece of chicken results in the inside being undercooked when the outside is ready.

Chicken is also sold in dismembered pieces. Pieces may include quarters, or fourths of the chicken. A chicken is typically cut into two leg quarters and two breast quarters. Each quarter contains two of the commonly available pieces of chicken. A leg quarter contains the thigh, drumstick and a portion of the back; a leg has the back portion removed. A breast quarter contains the breast, wing and portion of the back; a breast has the back portion and wing removed. Pieces may be sold in packages of all of the same pieces, or in combination packages. Whole chicken cut up refers to either the entire bird cut into 8 individual pieces. (8-piece cut); or sometimes without the back. A 9-piece cut (usually for fast food restaurants) has the tip of the breast cut off before splitting.

Pick of the Chicken, or similar titles, refers to a package with only some of the chicken pieces. Typically the breasts, thighs, and legs without wings or back. Thighs and breasts are sold boneless and/or skinless. Dark meat (legs, drumsticks and thighs) pieces are typically cheaper than white meat pieces (breast, wings). Chicken livers and/ or gizzards are commonly available packaged separately. Other parts of the chicken, such as the neck, feet, combs, etc. are not widely available except in countries where they are in demand, or in cities that cater to ethnic groups who favour these parts.

There are many fast food restaurant chains on both a national and global scale that sell exclusively or primarily in poultry products including KFC (global), Red Rooster (Australia), Hector Chicken (Belgium) and CFC (Indonesia). Most of the products on the menu in such eateries are fried or breaded and are served with french fries.

Cooking

Raw chicken can be frozen for up to two years without significant changes in flavour or texture. Chicken is typically eaten cooked as when raw it often contains *Salmonella.* Raw or rare chicken dishes appear in Ethiopian cuisine and Japanese cuisine.

***Figure:** Marination of chicken for grilling.*

Chicken can be cooked in many ways. It can be made into sausages, skewered, put in salads, grilled, breaded and deep-fried, or used in various curries. There is significant variation in cooking methods amongst cultures. Historically common methods include roasting, baking, broasting, and frying. Today, chickens are frequently cooked by deep frying and prepared as fast foods such as fried chicken, chicken nuggets, chicken lollipops or buffalo wings. They are also often grilled for salads or tacos.

Chickens often come with labels such as "roaster", which suggest a method of cooking based on the type of chicken. While these labels are only suggestions, ones labelled for stew often do not do well when cooked with other methods.

Some chicken breast cuts and processed chicken breast products include the moniker "with Rib Meat." This is a misnomer, as it is the small piece of white meat that overlays the scapula, and is removed with the breast meat. The breast is cut from the chicken and sold as a solid cut, while the leftover breast and true rib meat is stripped from the bone through mechanical separation for use in chicken franks, for example. Breast meat is often sliced thinly and marketed as chicken slices, an easy filling for sandwiches. Often, the tenderloin (pectoralis minor) is marketed separately from the breast (pectoralis major). In the US, "tenders" can be either tenderloins or strips cut from the breast. Chicken bones are hazardous to health as they tend to break into sharp splinters when eaten, but they can be simmered with vegetables and herbs for hours or even days to make chicken stock.

In Asian countries it is possible to buy bones alone as they are very popular for making chicken soups, which are said to be healthy. In Australia the rib cages and backs of chickens after the other cuts have been removed are frequently sold cheaply in supermarket delicatessen sections as either "chicken frames" or "chicken carcasses" and are purchased for soup or stock purposes.

Freezing

Raw chicken maintains its quality longer in the freezer as compared to when having been cooked because moisture is lost during cooking. There is little change in nutrient value of chicken during freezer storage. For optimal quality, however, a maximal storage time in the freezer of 12 months is recommended for uncooked whole chicken, 9 months for uncooked chicken parts, 3 to 4 months for uncooked chicken giblets, and 4 months for cooked chicken. Freezing doesn't usually cause colour changes in poultry, but the bones and the meat near them can become dark. This bone darkening results when pigment seeps through the porous bones of young poultry into the surrounding tissues when the poultry meat is frozen and thawed. It is safe to freeze chicken directly in its original packaging, however this type of wrap is permeable to air and quality may diminish over time.

Therefore, for prolonged storage, it is recommended to overwrap these packages. It is recommended to freeze unopened vacuum packages as is. If a package has accidentally been torn or has opened while food is in the freezer, the food is still safe to use, but it is still recommended to overwrap or rewrap it. Chicken should be from other foods, so if they begin to thaw, their juices won't drip onto other foods.

If previously frozen chicken is purchased at a retail store, it can be refrozen if it has been handled properly. Chicken can be cooked or reheated from the frozen state, but it will take approximately one and a half times as long to cook, and any wrapping or absorbent paper should be discarded.

Use of Roxarsone in Chicken Production

In many factory farms, chickens are routinely administered with the feed additive Roxarsone, a relatively benign organoarsenic compound which partially decomposes into inorganic arsenic compounds in the flesh of chickens, and in their feces, which are often used as a fertilizer.

The compound is used to control stomach pathogens and promote growth. A *Consumer Reports* study in 2004 reported finding "no detectable arsenic in our samples of muscle" but found "A few of our chicken-liver samples has an amount that according to EPA standards could cause neurological problems in a child who ate 2 ounces of cooked liver per week or in an adult who ate 5.5 ounces per week."

However, the amounts found in these livers averaged to 460 part per billion; an amount still less than the 2,000 parts per billion limit set by the FDA. The FDA has found there to be no significant impact for the use of Roxarsone in the production of chicken, turkey or swine.

Chickens as Pets

Chickens can be tamed by hand feeding and simply by being handled. Some people are afraid that roosters will become aggressive, but this problem can easily be avoided if the rooster is handled right. Breeds such as Silkies and many bantams are generally docile, making them ideal pets for owners with small children. Some cities in the United States allow chickens as pets but others ban them. Some may only ban roosters due to the crowing. City ordinances, zoning regulations or health boards may determine whether chickens may be kept. A general requirement is that the birds be confined to the owner's property, not allowed to roam freely. There may be restrictions on the size of the property or how far from human dwellings a coop may be located, etc.

Figure: *Baby chickens, dyed unnatural colours, sold as pets at a Market in Oaxaca, Mexico.*

The so called "urban hen movement" harks back to the days when chicken keeping was much more common, and involves the keeping of small groups of hens in areas where they may not be expected, such as closely populated cities and suburban areas. In Asia, chickens with striking plumage have long been kept for ornamental purposes, including feather-footed varieties such as the Cochin and the Silkie from China, and the extremely long-tailed Phoenix from Japan. Asian ornamental varieties were imported into the United States and Great Britain in the late 1800s. Distinctive American varieties of chickens have been developed from these Asian breeds. Poultry fanciers began keeping these ornamental birds for exhibition, a practice that continues today. Individuals in rural communities commonly keep chickens for both ornamental and practical value.

HousingA chicken coop is a housing where chickens are kept. Inside there will often be nest boxes for egg laying along with perches on which the birds can sleep. Backyard coops are small and fenced, often with chicken wire, allowing chickens an area to roam, peck and hunt insects. Chicken tractors are floorless coops which can be dragged about a yard. Some backyard chickens are allowed to free range, sleeping in coops. Urban chicken keeping has led to manufactured chicken coops such as the Eglu, which are designed for tight spaces and have a tidy look. Chicken waterers and feeders are an important part of keeping chickens as pets. There are hanging waterers/feeders, nipple waterers and nipple cup waterers.

Show Chickens

Chicken shows may be found both at county fairs and sanctioned shows hosted by regional poultry clubs. 4,000 or more birds may be entered in some shows. The Poultry Club of Great Britain sanctions poultry shows in Britain while the American Poultry Association and the American Bantam Association do likewise in America. Such organisations also work with poultry breed clubs of people who have interests in specific breeds of chickens.

Free-range Eggs

Free-range eggs are eggs produced using birds that are permitted to roam freely within a farmyard, a shed or a chicken coop. This is different from factory-farmed birds that are typically enclosed in battery cages. The term "free-range" may be used differently depending on the country and its laws. Legal definition or even nonexistent depending on the country. For example, the U.S. Department of

Agriculture requires only that the bird spends part of its time outside, and allows egg producers to freely label these eggs as free-range. Many producers will label their eggs as *cage-free* in addition to or instead of *free-range*. Recently, US egg labels have expanded to include the term "barn-roaming," to more accurately describe the source of those eggs that are laid by chickens who do not range freely but are confined to a barn instead of a more restrictive cage.

Cage-free egg production includes barns, free-range and organic systems. In the UK, free-range systems are the most popular of the non-cage alternatives, accounting for around 28% of all eggs, compared to 4% in barns and 6% organic. In free-range systems, hens are housed to a similar standard as the barn or aviary. In addition, they have constant daytime access to an outside range with vegetation. In the EU each hen must have at least 4 square metres of space.

Figure: *Free-range hens.*

Non-cagegunjan systems may be single or multi-tier (up to four levels), with or without outdoor access. Indoor non-cage systems are also referred to as aviaries (for systems with multiple tiers) or barn systems. The Laying Hens Directive stipulates that from 1 January 2007 (1 January 2002 for newly-built or rebuilt systems), non-cage systems must provide the following:

- A maximum stocking density of 9 birds/m^2 of "usable" space (units in production on or before 3 August 1999 may continue with a stocking density up to 12 birds/m^2 until 31 December 2011)

- If more than one level is used, a height of at least 45 cm between the levels
- One nest for every seven hens (or 1m2 of nest space for every 120 hens if group nests are used)
- Litter (e.g., wood shavings) covering at least one-third of the floor surface, providing at least 250 cm^2 of littered area per hen
- 15 cm of perching space per hen.

In addition to these requirements, free-range systems must also provide the following:

- One hectare of outdoor range for every 2500 hens (equivalent to 4 m^2 per hen; at least 2.5 m^2 per hen must be available at any one time if rotation of the outdoor range is practised)
- Continuous access during the day to this open-air range, which must be "mainly covered with vegetation"
- Several popholes extending along the entire length of the building, providing at least 2 m of opening for every 1000 hens.

Case studies of free-range systems for laying hens across the EU, carried out by Compassion in World Farming, demonstrate how breed choice and preventive management practices can enable farmers to successfully use non beak-trimmed birds.

Cost

Based on data in the European Commission's socioeconomic report, it costs €0.66 to produce 12 battery eggs, €0.82 to produce 12 barn eggs and €0.98 to produce 12 free-range eggs. So 12 free-range eggs cost €0.32 more to produce than 12 battery eggs, and 12 barn eggs cost €0.16 more to produce than 12 battery eggs. This means that one free-range egg costs 2.6 Eurocents more to produce than a battery egg, and a barn egg costs 1.3 Eurocents more to produce than a battery egg.

The Commission's report concludes that, if costs were to increase by 20%, which it says is the type of percentage increase in terms of variable costs that producers are likely to face as a result of switching to free-range, the industry will potentially suffer a loss of producer surplus of €354 million (EU-25). This appears to be a substantial sum. If, however, this increased cost were borne not by farmers but by consumers paying a little extra for eggs, each EU citizen would only have to pay less than €1 extra per year, as the human population of the EU-25 is around 460 million.

The margins achieved by producers for barn and free-range eggs are appreciably higher than those available for battery eggs. The Commission's socioeconomic report shows that margins for free-range eggs are around twice as high as those for battery eggs.

Retailers

A number of major retailers already have an express policy of only selling free-range eggs or of not selling battery eggs. Some retailers apply this policy not just to shell eggs but also to eggs used in baked goods and processed products such as ready-made meals, quiches, and ice cream. In the UK, The Co-Operative and Marks & Spencer sells only free-range shell eggs and uses only free-range eggs in their entire range of baked goods, processed products, and ready-made meals. Waitrose sells only non-cage shell eggs, and uses only free-range eggs in their processed products and ready-made meals.

As of 1 January 2007 (with one minor exception), all Austrian supermarkets no longer sell battery eggs. Many retailers in the Netherlands, including Albert Heijn and Schuitema (subsidiaries of Ahold), Laurus (including Edah, Konmar and Super de Boer), Dirk van den Broek (including Bas van der Heijden and Digros), Aldi and Lidl sell only free-range shell eggs. Three Belgian supermarkets: Makro, Colruyt and Lidl, no longer sell battery eggs. The Commission's report states that Sweden's move away from conventional battery cages has been aided by the decision by the four largest retailers (who between them account for 98-99% of the Swedish retail market) to stop stocking conventional battery eggs.

Misconceptions

Free range does not imply in any way that the hens were fed any differently than on normal commercial farms. The label "free roaming" does not describe feed supplies, which means that free-range hens can be fed the same animal-derived by products or GMO crops as in other non-organic farms. This is also the main reason why free-range eggs are cheaper than organic eggs.

Consumers of free-range eggs want eggs from hens that are kept under traditional low-density, free-range conditions. Critics of EU-style free-range regulations point out that commercial free-range egg farming, in general, does not live up to these consumer requirements, since the regulations allow the use of yarding rather than free range.

Yarding combines a high-density poultry house with an attached fenced yard, and both its methods and results are closer to high-density confinement than true free range. Free-range eggs may be broader, and have more of an orange colour to their yolks due to the abundance of greens and insects in the diet of the birds. An orange yolk is, however, no guarantee that an egg was produced by a free-range hen. Feed additives such as marigold petal meal, dried algae, or alfalfa meal can be used to colour the yolks.

Nutritional Content

Studies suggest the nutritional content of eggs from genuine free-range hens (hens that forage daily on a grass range) is superior to that of eggs produced by conventional means.

These studies report higher levels of Omega 3 and Vitamins A and E, and lower levels of total fat, saturated fat, cholesterol, and Omega 6. An egg study comparing free-range eggs to the U.S. Department of Agriculture (USDA) nutrient data for commercial eggs. The findings showed that free-range chicken eggs produced the following results:

- 1/3 less cholesterol
- 1/4 less saturated fat
- 2/3 more vitamin A
- 2 times more omega-3 fatty acids
- 3 times more vitamin E
- 7 times more beta-carotene.

A study by the U.S. Department of Agriculture in 2010 using industry standard measures of shell strength, the height of egg whites (Haugh unit) and their protein and crude fat content determined that there were no nutritional benefits to free-range eggs when compared to factory eggs. However the study did not measure the types of fat in the eggs nor did it measure differences in vitamin and essential fatty acid content.

Organic Egg Production

Differences between "Free Range" and "Organic"

Significant differences cover feed, medication, and animal welfare. Organic hens are fed organic feed; it is prohibited to feed animal by products or GMO crops - which is not disallowed in free range environments; no antibiotics allowed except in emergencies (in free range, it is up to the farmer, but the same levels of antibiotics as conventional farming is allowed); required animal welfare standards in organic farms, which can improve the quality of both the eggs and the meat - low stress levels lead to superior quality of animal products, a fact long known and used in the production of the famous Wagyu beef. Look for labels that say both "Organic" and "Free-Range". Cage free only means that they are not in a cage, but are usually still confined in barns at high population densities.

Living Conditions

***Figure:** At an organic farm (in Bruthen, Victoria) chickens sometimes end up laying eggs somewhere other than the owner expected*

Organic Feed

Organic feed is grown by certified organic farmers. To become a certified organic farmer, the crop must be free of genetically modified organisms (GMOs). The crop must be free of GMOs and synthetic fertilization for three years before it can be certified for organic usage. If the crop is contaminated by cross-fertilization, the crop is rendered useless for organic grading. Finally, there can be no animal by-products fed to the poultry. In the United States, "organic" egg production means that the flock may not live in cages and must have access to the outdoors. Organic egg producers in the United States do not always grant meaningful outdoor access to their organic laying hens; most industrial-scale organic egg producers generally build small wood or concrete porches attached to the henhouses, which passes as "outdoor access."

Antibiotics

Organic egg producers cannot feed low-level antibiotics to the poultry. Antibiotics are only allowed during an outbreak of infection or disease.

Moulting

Some farms induce molting in their flocks to affect egg production. In organic egg farms, the birds are allowed to go into a natural moult but are not induced.

Animal Welfare

Similar to all other forms of egg production in the United States, organic production is also regulated by animal welfare audit system. Mistreatment of the chickens could potentially lead a farmer to losing their organic certification. Thus some of the arguments from animal activist organisations that egg production is cruel and inhumane do not necessarily apply to organic raised hens. On the other hand, male chicks who are born on organic or free-range egg farms are still discarded, by the use of lethal gas, because they do not produce eggs. The American Organic Standards still allow beak trimming as a means to lessen injury to the birds.

Pastured Poultry

Pastured poultry is a sustainable agriculture technique that calls for the raising of laying chickens, meat chickens (broilers), and/or turkeys on pasture, as opposed to indoor confinement. Humane treatment, the perceived health benefits of pastured poultry, in addition to superior texture and flavour, are causing an increase in demand for such products. Joel Salatin of Swoope, Virginia, helped to reintroduce the technique at Polyface Farm, and wrote his book *Pastured Poultry Profits* to spread the idea to other farmers. Andy Lee and Herman Beck-Chenoweth expanded on Salatin's techniques, and created some of their own. The American Pastured Poultry Producers' Association (APPPA) was formed to promote pastured poultry. Its membership consists largely of pastured poultry farmers.

Though pasture feeding improves the nutritive quality of ruminant meats, the effect of pasture feeding on poultry meat composition is not well established. One trial showed low impact of pasture feeding on vitamin E and fatty acid composition.

The pens that house the fowl can be made from wood and scrap metal or out of PVC pipe and white tarps. Pastured poultry is also gaining popularity because it helps the farmer, through reducing capital costs, and increasing pasture fertility. It is very well suited for incorporation within a system of managed intensive grasing. Pastured Poultry is not limited to chickens and turkeys. It includes a variety of other birds, including ducks, geese and exotics in the poultry family.

Free-Range Poultry

Herman Beck-Chenoweth reintroduced the free-range system that was the most popular way to raise poultry in the U.S. from the 1930s through the 1960s. The system allows birds to range freely during the day and be safely sequestered on secure skid houses over night. The addition of a guard animal, such as an Komondor or Anotolian Shepperd dog, controls predators. In the Modern American Free-Range Poultry Production System birds are much less crowded and freer to practice normal bird behaviour than in any other pasture based system. Although frequently listed as a "pasture" method, free-"range" refers to the length of the forage. Cows graze "pasture" which is forage over six inches long. "Range" refers to short forage of 2-4 inches. Free-Range is a very sustainable production system that improves the farmer's soil and produces poultry with strong bones and meat with good "mouthfeel". Combined with proper aging after slaughter the meat is tender and flavourful.

***Figure:** A chicken tractor in use as part of a pastured poultry system*

Eggmobiles

As part of their pastured poultry process, some farmers use a mobile house that houses egg laying hens. This is called an eggmobile. Popularised by Joel Salatin, this model has been adapted across the United States. The concept is to allow chickens to move two to three days behind cows, which are rotationally grasing. The chickens scratch through cow manure, harvesting their own larvae and insects, spreading out manure and helping to reduce the fly population on the cows. The eggmobile contains nesting boxes that the hens use to lay eggs. The nesting boxes can be made of many materials, such as this design at Nature's Harmony Farm in Georgia which uses milk crates.

The foraging helps the hens to produce eggs with an "orangish" yolk and a thicker albumen. This is owed to the beta carotene that the grass provides to the eggs. A test done by Mother Earth News revealed that eggs laid by hens foraging on pasture have shown:

- 1/3 less cholesterol
- 1/4 less saturated fat
- 2/3 more vitamin A
- 2 times more omega-3 fatty acids
- 3 times more vitamin E
- 7 times more beta carotene.

National Chicken Council

The National Chicken Council

(NCC), based in Washington, D.C., is the non-profit trade association representing the United States chicken industry. NCC is a full-service trade association that promotes and protects the interests of the chicken industry and is the industry's voice before the United States Congress and United States federal agencies. NCC member companies include chicken producer/processors, poultry distributors, and allied industry firms. The producer/processors account for approximately 95 percent of the chickens produced in the United States. The NCC's website contains a video about biosecurity in the poultry industry and links to its own avian influenza page, and additional references regarding avian influenza on its "About the Industry" page. The industry also is sponsoring a separate website with news, information, and resource links.

Modern Agriculture and Farming

Welcome to online farming guide. Bangladesh is an agricultural country. The economy of our country is mostly depend of agriculture. A major part of our population are directly or indirectly involved with agriculture. Income source of our people are limited. So they raise, poultry, ducks, dairy, fish etc. to earn some extra income. So, in this site we have tried to show you some possible way of earning. Not only for Bangladesh but also about Indian agriculture and almost all Asian agricultural countries.

We have launched in the concept of modern agriculture. We don't like traditional agriculture. Modern agriculture has changed the total agricultural process. So modern agricultural revolution is a must. To be success we need the proper, correct and experimental information based on modern technology. Which can make our dream true fast. Farmers to our country are doing their farming by the knowledge and experience which they obtain from their relatives.

Modern technology in agriculture sector did not make any progress. Many eligible and famous researchers in our country are constantly researching and getting good results. But the main issue is that their results are not being published well. In some cases, the results of the research are restricted to journals and publication. We want to share the achievement of our researchers to the mass people.

There are many agricultural university in our country to improve our agricultural condition. Many types of agricultural journals available in every parts of the worlds. Modern meditation, ideas, technology in agriculture and seeds in the land, to ensure the maximum use of fertilizer, raising poultry, ducks, dairy, goats, fish and many more, we have tried our best to collect all about this in our site. Due to improve the socioeconomic condition of our people we encourage land based farming as well as cattle, birds and the fish farming. Not only the well production, we have to be ensure about the market of our products.

2

Poultry Feeds and Feeding

Definitions and Terms Used

Nutrition: The process of digesting, absorbing and converting food into tissue and energy. Also, the study of this process.

Nutrient: A substance that can be used as food. Some people use "food" for nutrients eaten by people and "feed" or "feedstuff" for animals.

Digestion: The process of changing food to a form that can be absorbed from the digestive tract by the body tissues (mainly the intestines). In the digestive tract this is done by enzymes and other material produced by the digestive tract which break down the food into small, simple (molecular) components.

Digestion by microorganisms: Some bacteria and protozoa produce enzymes and other material that break down cellulose, and fibre, etc. that non-herbivores (non-grass eaters) like chickens cannot digest. There is some digestion by microorganisms in the cecum of chickens. Enzymes produced by bacteria can be added to some feeds to improve digestion and absorption.

Metabolism: A chemical reaction that takes place in the tissues and organs of the body in which the food that has been digested and absorbed is changed either into energy or building blocks for the body. Energy is the power produced by the food. It is the fuel, like petrol for cars, on which the body runs and, like burning wood or petrol, metabolism requires oxygen and produces heat and waste material (carbon dioxide and water).

Basal metabolism, basal metabolic rate (BMR): The amount of energy (fuel) required to keep the body alive and operating without activity, growth or production.

Energy: Energy is the amount of power produced when food is metabolised. Energy is measured in heat units (calories or joules). In nutrition the kilocalorie (kcal) equals 1000 gram calories. A gram calorie is the heat required to raise 1 gram of water 1BC (14.5 to 15.5BC) (1 kcal equals 4.184 kjoules).

Metabolisable energy (ME): The amount of energy available to be used for maintenance, for production of body tissue (for growth and replacement), activity and egg production, when a food material or feed is eaten. It includes the heat lost during metabolism. The ME of a feed ingredient (individual feed) may be used to indicate the nutritional value of that ingredient. Feed ingredients or feeds are rated as high or low energy. ME depends on the quality of the feed and on the % dry matter. Good maize (corn) at 85% dry matter (15% water) has an ME of 3300 kcal/kg. The ME for barley is 2700 kcal/kg. Fats may have an ME of 9000 kcal/kg. If chickens are fed a low energy feed they will eat more feed, if it is available, to get the required energy. In monogastric animals, like chickens, energy comes mainly from carbohydrates and fats since fibre containing cellulose cannot be digested.

Organic: Organic compounds are defined in nutrition as animal or plant material containing carbon.

Carbohydrates: Carbohydrates are organic compounds and a source of energy for poultry. Simple carbohydrates are made up from sugars: glucose from maize is a monosaccharide. Lactose from milk, sucrose from sugarcane or sugar beets and in smaller amounts in many plants, particularly in their fruits, seeds or roots are disaccharides. Complex carbohydrates (polysaccharides) are made up from combinations of sugars. Some, like starch (from cassava, maize, wheat, etc.) can be digested by chickens while others (cellulose, a structural carbohydrate) cannot because animals do not have the enzymes to break down (hydrolyse) cellulose so that it can be absorbed. This digestion is done by bacteria and protozoa in herbavores.

Fats (oils, lipids): Fats are a source of energy and in some cases fat soluble vitamins. Like carbohydrates, they are organic compounds made up from carbon, hydrogen and oxygen which form fatty acids. Because they are higher in hydrogen and lower in oxygen than carbohydrates, fats have a higher energy value (ME) than carbohydrates. Fats that are high in unsaturated fatty acid are liquid at room temperature and are called oils. Vegetable oils (canola oil, corn oil, cottonseed oil, soybean oil, etc.). Most essential fatty acids (required for health &

growth) can be produced by the chicken from other food; but linoleic acid must be present in the diet of chickens for proper growth and egg production.

Proteins: Proteins are organic compounds used to build the cells, tissues and organs for the body. They are made up from many amino acids some of which are essential for growth and production. Proteins contain about 16% nitrogen so the amount of protein (crude protein, not the digestible or usable protein) in a feed can be estimated by measuring the nitrogen content. Excess protein (above what is required for growth or production) is used as energy and the nitrogen excreted in the urine. During digestion protein is broken down into individual amino acids for absorption. In the body these are reassembled to make body tissue or egg protein. Birds can make some amino acids from other protein but many amino acids are essential, that is they must be present in the diet. Proteins from animal sources (meat, milk, insects, larvae, etc.) contain the essential amino acids. Most plants are low in protein and vary widely in essential amino acid content. Most are low in one or more essential amino acids. Some legume plant seeds such as beans are high in protein.

Vitamins: This term describes a variety of essential nutrients that are not similar to one another, except that they are essential in the diet, but only required in very small amounts. They are used as metabolic regulators. They are either water soluble or fat soluble. Vitamins or their precursors are present in small, but variable amounts in some feeds. Most vitamins are manufactured (synthetic) for use in commercial feed. Precursors of vitamin D3 can be produced in the skin by sunlight. Vitamin C is produced by bacteria in the intestine in chickens.

Minerals. Minerals are Not Organic: They are chemical elements. Those required in small but significant amounts (calcium, phosphorus, sodium, potassium, magnesium and chlorine) are major minerals. Essential trace minerals are iodine, cobalt (as cobolamin), iron, copper, zinc, selenium and molybdenum. Most minerals must be added to the diet for good growth and egg production. Laying hens need a ration with 3% calcium to make egg shells. This can be supplied free choice as small pieces of bone-meal, coral, sea-shell or limestone. Phosphorus in plants is present as phytate phosphorus and is only partially available unless phytase enzyme is added to the feed.

Water: Water is an essential nutrient. Water is the major part of animal tissues and all body functions require water. A source of clean, cool water should always be available free choice.

Antibiotics: Antibiotics are drugs made from bacteria or fungi that are used as medicine to prevent or treat bacterial disease. They are sometimes used as growth stimulants in chickens.

Probiotics: Probiotics are bacteria or products produced by bacteria that encourage the growth of "good" bacteria (those that prevent the attachment and or growth of disease causing bacteria) in the intestine.

Chemotherapeutics are chemical compounds used as medicines.

Vaccines are preparations of live organisms, (mainly viruses) used to produce immunity (defence) against disease causing viruses. They stimulate antibody production by the chicken to protect against the virus material in the vaccine. Because they are live and must multiply in the body to be effective, they can be used as a spray, in water, eyedrop, wing web injection or subcutaneous injection.

Bacterins are preparations of killed organisms (bacteria, mycoplasma, virus, etc.) for subcutaneous or intramuscular injection. They also stimulate antibody production.

Antibodies are very small particles produced by lymphocytes (type of blood cell) that circulate in the blood to help the body defence systems stop infection by specific virus or bacteria that had stimulated the lymphocytes to produce the antibody. They can also be passed in the egg from the hen to the chick (maternal antibody) and protect the chick for 1-3 weeks.

Cereal Grains: Cereals are the grains grown specifically from human and animal food. They include wheat, millet, rice, maize (corn), sorghum (milo, kafir or guinea corn), barley, oats, etc. Grains provide the main source of energy in commercial poultry feed. Protein in cereals are low (8 to 12%) and the quality (level of essential amino acids) is poor.

By-Products: By-products are the parts of a grain, oilseed, or animal that is being prepared for human food, that is not used for human food. Examples are wheat or rice bran or animal viscerae that may be fed to chickens.

Complete Feed: A prepared feed that contains all of the nutrients for the best growth or egg production for the flock for which it is being used is called a complete feed. These feeds are usually prepared commercially by a feed manufacturer. Broiler chickens receive high protein (21 to 23%) starter for the first 14 to 21 days, grower to day 28 or longer and finisher, that is lower in protein and higher in energy, until ready for market. Supplement or concentrate is a prepared feed that is intended to be added to or mixed with other feed material to improve the nutrient balance of the final feed.

Premix: A premix may contain a variety of specific ingredients such as vitamins, trace minerals, amino acids, or medicine that is to be mixed with other feed to supply essential or important elements that may be missing. Essential nutrients are sometimes added to the water if they are water-soluble.

Antinutrients: Some feed ingredients and potential feeds contain factors that inhibit the digestive process causing reduced growth, diarrhea or pasting. They limit the amount of some feed ingredients that can be added to the final feed. The antinutrition factors in some feed material such as beans can be destroyed by heat (cooking).

Phytotoxins are toxic or poison substances found in plants used as feed for chickens. The toxic material can be in the seed (castor bean) leaf or stem or in the root or tuber (cassava). At low levels some might only reduce growth rate or have no effect. At higher doses they might cause illness or death. Toxic weed seeds may contaminate grain that is being harvested making that feed toxic.

Mycotoxins are poisons produced by moulds growing in food material. Various fungi produce different toxins. Aspergillus fungi produce aflatoxin, one of the most serious in hot climates. The fungi can grow in crops in the field, in seeds in storage or in prepared feed for chickens.

Nutritional Requirements of Poultry

Poultry require carbohydrates, fats, protein, vitamins, minerals, and water. Most nutrients provide both energy (as carbohydrate or fat) and protein. Feedstuffs are usually classed as being primarily a source of energy or protein. Fats and oils do not provide protein. Energy and protein are more efficient if they are available in the proper ratio. Excess protein is used as energy with increased waste excretion. Growing chickens should have 16-24% protein; growing turkeys, 24-28% protein; for egg production 15-17% protein; for maintenance (no growth or production) 10-12% protein. The highest requirement is in the first 2 to 3 weeks and is higher in young leghorns than broilers.

The protein level can be lower if the essential amino acids are all present at the proper level. The vitamin and mineral levels are higher for growing birds and for egg production. More rapid growth requires more added vitamins and minerals than slow growth. In slow growing birds more of some of these essential vitamins and minerals are available from the feed ingredients because the amount of feed required for basal metabolism is greater.

Some feed and water is required just for survival. The BMR varies directly with the weight of the bird. If the feed energy required to supply the basic energy need is not available, the bird will lose weight, get thin and die. Some protein, fat, vitamins and minerals are also required for basic survival. Nutrients available above the basic level can be used for growth in young birds and egg production in adults. Excess energy is converted to fat and stored in the body. It may be used when energy from feed is lacking as in the hump in camels and tail fat in fat-tailed sheep.

The rate of growth in young birds or egg production in adults is controlled by:

a) the genetic potential (commercial broilers compared to layers)
b) the amount of energy and protein (and protein quality) available above the level for the BMR. The feed required to maintain BMR does not add to growth or production. The maintenance requirement for a slow-growing broiler is the same as a fast-growing broiler of the same weight. The feed maintenance requirement for a hen laying 2 eggs a week is the same as a hen of the same weight laying 6 eggs a week. Sufficient vitamins and minerals must also be available.

If the temperature is below the birds comfort level, additional energy is required to provide internal heat by increased metabolism to maintain body temperature. Activity also increases metabolism and birds that are allowed to run outside or that scavenge for their feed have a higher requirement for energy.

Chickens that get all their nutrients from scavenging may eat an excess of protein, if insects, worms, larvae, etc. are available. They might benefit most from supplemental feeding of energy in the form of carbohydrate (cereal grains, etc.). Fenced or backyard poultry fed household or garden waste may lack both energy and protein for good growth or egg production.

Laying hens should have calcium available free-choice, even if calcium is being added to feed. Chickens that eat whole seeds, grains, vegetable material or fibre must have insoluble grit (granite grit) or small stones in their gizzard to grind the hard or fibre material. Birds that forage pick up their own grit. Poultry kept indoors must have grit supplied. Small chicks require small stones 2 to 4 mm. Hens .5 to 1.5 cm. Limestone particles are not satisfactory as grit. The amino acids that growing poultry require are: arganine, cystine, lysine, methionine, tryptophan, histidine, isoleucine, leucine, phenylalanine, threonine, tyrosine, valine. The following vitamins should be added to a prepared feed: vitamins A, D3, B12, riboflavin, pantothenic acid, niacin, folicin (folic acid), thiamine and biotin.

Commercial Nutrient Sources

In countries where complete feeds are prepared by feed manufacturers, cereal grains and oilseed meal are the major ingredients in the feed. Depending on cost, availability and age or type of poultry or water fowl, some of the following products are also frequently used: by-products of cereal grains processed for human use, animal processing waste, fishmeal, waste oil from restaurants (restaurant grease), yeast, alfalfa meal, distillery or milk factory products, etc. Vitamins and mineral are also added in the correct proportions. The ingredients are ground (if required), mixed and may be pelleted.

Ingredients most frequently used:

- for energy: maize (corn) up to 65%
- sorghum (milo) up to 45%
- wheat up to 25%; with enzyme up to 45%
- wheat by-products (bran, shorts, screenings) up to 15%
- rice up to 15%, rice by-products (bran, polishings) up to 15%
- barley up to 15%; with enzyme up to 35%
- molasses, up to 5% after 2 weeks
- for protein: soybean meal up to 30%
- soybeans up to 15% (heated to remove antinutrients)
- canola meal or whole seed up to 10%
- corn gluten up to 15%
- peas, lupin, flax up to 10%; flax (linseed) meal - 15 % (20% in layers)
- safflower meal, sunflower meal up to 10%
- meat meal, fish meal up to 10%
- blood meal, feather meal up to 2%
- Fats and Oils: tallow, lard, coconut oil, palm oil, after 3 weeks up to 5%

- poultry fat, fish oil, restaurant grease up to 3%, after 3 weeks 6%

Minerals: Calcium can be added as limestone, as prepared products such as dicalcium phosphate or as oyster or other marine shell. Growing birds require about 0.9% calcium in the diet. Hens in lay require 3 to 4% calcium in the diet so that calcium is available for egg shell production.

Phosphorus can be added from manufactured products such as dicalcium or monocalcium phosphate. Phytate phosphorus (an organic phosphorus) is present in small but varying amounts in plants feed ingredients. This phytate phosphorus is only 25-50% available unless phytase enzyme is added to the ration. Growing birds require about 0.4% phosphorus, adults about 0.3%. Calcium and phosphorus levels must be in balance since high levels of one cause deficiency of the other.

Sodium and chlorine are usually added as salt (NaCl) at about 0.25% of the ration (2.5 kg NaCl/1000 kg).

Magnesium and potassium are usually present in adequate amounts in the feed ingredients. Some ingredients such as soymeal and molasses are high in potassium. Minerals may be present in high or moderate levels in animal by-products. If animal by-products are being fed the minerals present in this feed should be used in the calculation of the amount to be added. High levels of sodium are toxic to birds. Saline water is also toxic for birds.

Trace minerals and vitamins are required in small quantities. They are not usually added as separate ingredients. They are usually bought as a prepared package or a premix. A specified amount of this premix can then be added to each tonne (1000 kg) of feed. To make sure of an even distribution in the final feed the vitamin-mineral premix may be mixed with 25-50 kg of cornmeal or other large particle ingredient that will help ensure an even mix. This is then added to the larger mixer after the cereal and protein portions have been put in. The calcium, phosphorus and salt should never be added to the vitamin premix. The mixer should be grounded.

Alternative Feeds

There are many vegetable and animal products that can be used as feed ingredients for poultry. Poultry cannot digest fibre found in roughage, seeds, grains and fruits. Products that are high in fibre can make up only a small percent of the feed. Since many plant products also contain antinutrients the amount of these feed stuffs that can be fed is also limited. Products that contain a high percent of water (kitchen and garden waste, weeds, tree and plant leaves, fruit, worms, larvae, fresh fish and animal waste, water animals and insects, milk products, etc.) can be given free-choice. What is not eaten must be removed so the chickens will not eat mouldy or spoiled material. Most feeds can be sun-dried to reduce the water content. They can then be ground into meal to be mixed with other feed, or cut up into chips that the birds can eat.

Many fresh (undried) foods such as garden and kitchen waste, fruit, plants and vines can be fed free choice. The best way to preserve foodstuffs is by sun-drying. It is important to avoid spoilage. Some types of foods can be preserved by anaerobic (no air) fermentation (making silage) where the fresh food ferments and becomes acid. The acidity prevents the growth of spoilage organisms. With plant material this may be done in a large plastic bag kept in the shade. Milk may be preserved for short periods by allowing it to sour (go acid) or by adding acid material. Seeds and grains that contain antinutrients and toxins must be used with care since levels of antinutrients and toxins vary. Several varieties of raw or unprocessed peas and bean should not be used together (don't use 5% of each) because they contain similar antinutrients and toxins.

Proteins are fed for the nitrogen and amino acid content. Some amino acids are essential and must be present in the feed or added as synthetic amino acids. Poultry cannot make them from other amino acids or nitrogen. Feeding a variety of protein sources may supply more of the essential amino acids.

Poultry require both sodium and chlorine. There is some sodium and chlorine in feedstuffs, but both must be added to the feed, usually as salt. The recommended level is 0.25 to 0.3% added salt (2.5 to 3 kg per tonne). If feed ingredients or drinking water contains salt, the amount of added salt must be reduced. Poultry can be poisoned by the sodium in salt and young chickens are very susceptible to high sodium. Chicks are frequently poisoned by salt in the drinking water. Water for young chicks should not contain salt and many chicks die from salt poisoning at levels below 0.1% (0.9% is the level found in animal tissue). Chicks can be poisoned by eating salted fish and particularly by the brine used to preserve fish.

Young chicks can also be poisoned by high calcium levels in the feed. Feed for layers contains 3 to 4% calcium. That level can cause kidney disease in young chicks.

The Importance of Water

Water is an essential nutrient. All body functions and processes require water. Water is the material in which all other nutrients are carried in solution. The tissues and cells in the body are made up mainly of water. Water is also present between cells so that material can be moved between cells. Dry feed contains only 10-15% water and poultry need about twice as much water as feed (2 gms of water for each gm of dried feed).

Water should be clean and free from chemicals and minerals. It should not contain harmful parasites or bacteria. Water should be easily available free-choice. If water consumption is restricted, growing chickens will grow more slowly. Broilers that get only 2/3 or 1/2 of the amount of water they want will eat less feed and grow at only 3/4 of the expected rate. In cool climates water restriction is used to slow growth in breeders or to reduce the ascites syndrome. If water is not freely and easily available adults will lay fewer eggs and may suffer from kidney disease (as do people that do not consume sufficient water). Layers that go without water for a day or two will stop laying and may take 2-3 weeks to recover. Water in a trough or cup must be deep enough to allow the bird to drink properly. A depth of 2 cm is adequate, unless poor beak trimming has resulted in long lower beaks. The drinkers must be easy to reach and housed poultry must not have to walk too far to get water; not more than 2 metres for broilers and 4 meters for adults.

In hot climates water availability is even more important. The amount of water required becomes much higher as the temperature rises above 25BC. Birds cool themselves by water evaporation from their respiratory system and lose water when they mouth breath. In hot climates cool water is better than warm water. For large groups of poultry the water pipes to the poultry sheds should be 40 to 60 cm underground to cool the water as it flows to the chickens. Above ground pipes and emergency tanks should be shaded from the sun. To keep the water in emergency tanks fresh it can be used for 3 to 5 hours in the early morning. Chickens will move to cooler parts of the pen when the temperature is high, particularly if there is a breeze or fans. It is important to have water easily available in these areas because the birds may not go back to hot areas to drink.

Nipple drinkers help conserve water and avoid spillage, but chickens drink less water from nipples. Nipple drinkers require careful management. The pipes must be level, at the correct height (at eye level for chicks, just above eye level for older birds) and the water pressure must not be too high for the birds to operate the nipple. The water column that controls pressure should be about the same height as the chickens using the system (6 cm for a 6 cm chick).

Newly hatched chicks should have water before or at the same time as they receive their first feed. It must not be in an open pan or trough that the chicks can get into. Wet chicks lose body heat quickly and may die. Water jugs inverted in pans with a narrow water source are satisfactory as are raised narrow troughs, bell-type drinkers and nipple drinkers. Water jugs, troughs, etc. must be cleaned at least once a day.

Water Problems

Water containing physical particles or dissolved mineral material may interfere with automatic drinkers. Particulate material and iron may be removed by a sand filter and settling tank but dissolved minerals and algae growth in the system are more difficult to control. Regular cleaning and flushing may be necessary. Copper sulphate solution may control fungi and algae growth, but it can be toxic for poultry. Poultry may be poisoned by minerals or chemicals in the water. The sodium in saline (NaCl) water is the most frequent problem. Young birds are very susceptible to sodium toxicity. Sodium above 500 ppm (0.05%) in the drinking water may cause death in young chickens and turkeys depending on the sodium level in the feed. The young birds usually die from oedema and ascites syndrome. Wet droppings or diarrhea would also occur. Salt in the feed can be reduced to avoid disease from salt in the water up to about 1000 ppm (0.1%) of sodium in the water. Saline water above 1000 ppm should not be used for broilers up to 21 days of age even if no salt is added to the feed. High levels of sodium will kill adult chickens by causing diarrhea and dehydration.

Sodium may be present in water as sodium sulphate (NaSO4) or sodium bicarbonate (NaHCO3). Sulphate may be present as magnesium sulphate (MgSO4) and cause diarrhea and death. All sources of sodium in the feed, whether added as salt or present in the feed ingredients, (particularly animal protein) and water are additive in causing sodium toxicity. If sodium is too high in the feed it will cause disease as it does in the water. Sodium, however, is an essential nutrient and poultry require some sodium to grow and produce eggs.

Nitrate in feed and water may be toxic at high levels but can reduce growth at 50 ppm in water. A variety of other naturally occurring minerals and chemicals may be present in water and may cause problems. Surface water may be contaminated by farm pesticides, fertilizer or by industrial chemicals.

Brooding and Growing Chicks and Poults

Hatched chicks should be lively, firm and healthy. They should be uniform in size and have a well-healed navel. They should be started as soon as possible after removal from the incubators. The sooner they are provided a warm area, feed and water, the better the rate of success. Chicks must not be chilled or overheated at any time. If chicks do not receive feed and water within 18 hours after hatching they should get water for 1-2 hours before they are given feed.

Systems of Brooding

All pens and equipment should be ready, tested and operated before the chicks arrive.

Natural Brooding: For small numbers, a broody hen can handle 14-15 chicks. Protection from predators and rain is important. Chicks should get some commercial ration or other feed for the first 3-4 weeks. Sour milk is good for young chicks. A little care over the first 3-4 weeks will often double or triple the survival rate.

Warm Room Brooding: In this system, the whole pen is held at a temperature of 30-32BC both day and night. The temperature is lowered about 1.5BC per week until the ambient temperature is reached, but should not go below 21BC until 6 to 8 weeks and not below 18BC until 10 to 15 weeks. Light should be over feed and water.

Circle Brooding: In this system, brooding areas of the pen under the heat reflector are heated to around 35BC by the use of reflectors over some form of artificial heat, such as gas heaters. Chicks are enclosed by cardboard chick guards, (not more than 400 chicks per circle) that allow them to move away from the heat source, but keep them close to feed and water. Light should attract chicks to the heat source. Temperature in the rest of the pen can be allowed to drop to 20 or 15BC. Heat may be supplied by hot water or hot air with the heat source usually oil, gas, coal, wood or electricity. For small flocks of up to 75 chicks, a heat lamp may be all that is required. Temperature readings should be taken 5 cm above the litter (or at chick level). Make sure the thermometer is shaded from radiant heat.

With circle brooding, the feed and water should be placed at the edge of the heat reflectors, which is 60 to 70 cm above the floor.

Comfortable chicks will be at the feeders and waterers and spread uniformly over the pen area. When chicks feel cold, they will crowd under the heat source or pile up.

If the pen is too warm, the chicks move away from the heat with wings spread and/or are panting.

Uncomfortable chicks will be noisy:

(a) Floor space - Allow 1m2 per 75-150 chicks under the heat source and at least 1m2 per 25-50 chicks within the chick guard. Many producers use a chick guard of corrugated cardboard 30 cm in height placed 2-3 m back from the heat source for the first 5 to 7 days. The guard should be removed when the birds can fly over it. Broilers at 6 weeks require 1m2 per 9 to 12 birds.

Leghorn chicks for egg production are often started in large (1 x 2m) or smaller growing cages with warm room brooding, or in hot climates with a heat source over each cage. Small growing cages may be about 50 x 50 cm with 24 chicks started in each. After 4 to 5 weeks, half the chicks are placed in another cage of equivalent size, to double the space per bird and split again to 6 per cage at 8 to 10 weeks (400 cm2 per bird).

(b) Feeders - Chicks should be given 5 (Leghorns) -8 (broilers) cm of feeding space. Feed is usually put in shallow boxes or egg flats on the floor or cage bottom for 4 to 6 days as well.

(c) Waterers - Good water is important. In large flocks, some form of automatic water system is usually installed. In a float controlled trough, 2 cm per bird is considered sufficient. One hanging fountain per 80 to 100 birds, one smaller cup per 50 birds, one nipple for 10-12 birds, will generally be adequate.

For small groups, three 4L or larger drinkers should be provided per 100 chicks, and more added as required. These should be cleaned and refilled daily. Feed and water should be within 1.5m of all chicks. Water should be clean and free from toxins and chemicals. Salt as NaCl or $NaS0_4$ is particularly dangerous in young chicks up to 21 days. Total sodium (Na+) in water should not be above 300 ppm for broilers and 600 ppm for Leghorns or local strains.

Feeding

For best results, a commercial ration is recommended. If home grain or other feed is available, it may be fed starting at day 8 in broilers and day 15 in Leghorns. Replace 5% of commercial ration with whole grain or other feeds and increase by 5% a week in broilers (5% every 14 days in Leghorns) to a maximum of 50%, or a commercial concentrate may be diluted with whole or ground grain or alternate feed stuffs. When grain is fed, supply insoluble grit sprinkled on the feed at least once per week. Don't overfeed grit.

Chickens raised for meat should be given all the feed they will eat (unless they are being raised at high altitude). Layers may have to be restricted to prevent them becoming too fat. Breeding flock must be monitored regularly and weighed weekly to make sure they are uniform and not overweight. Severe restriction may be required in broiler breeders. It is advisable to include an anticoccidial drug in the feed of birds grown on the floor. This should be fed continuously throughout the growing period up to market in broilers and to 12 weeks in breeders or Leghorns. Coccidia vaccines are also available.

Cannibalism (Feather Pulling - Vent Picking, Toe Picking, Etc.)

Birds are cannibalistic by nature and when confined may start picking.

Chicks should be beak-trimmed at the hatchery, or at the farm using an electric trimmer. Broilers should not need trimming but watch for signs of picking and beak-trim if necessary.

Other steps to consider are:

- decreasing the light intensity;
- decreasing pen temperature;
- provide more space;
- use red light bulbs or place red plastic over windows;
- feed whole oats if the flock is over 4 weeks of age;
- add 1 g salt per L of drinking water if flock is over 6 weeks of age.

Once started, cannibalism is difficult to control.

Trouble Shooting

- Chicks should be well spread out and at the drinkers and feeders
- If noisy and crowding the walls or chick guard they may be too warm
- If noisy and crowding the heat source they may be too cool or drafty

- If crowded around feeders or drinkers, check supply and space per bird.
- If chicks are huddling or piling in small lots they may be too cold, too hot or sick.
- Litter condition:
  - o too dry, may cause respiratory problems - spray litter to increase humidity.
  - o too wet, ammonia and coccidiosis or leg problems occur - check ventilation and bird floor space.
- Insufficient feeder or waterer space, disease, poor lighting, etc., will all affect uniformity.

Poultry Health Management for Commercial Poultry

Disease Prevention

Diseases caused by infection with a living microorganism such as bacteria, virus, mycoplasma, parasite, etc. are infectious diseases. Most infectious diseases are also contagious, that is, they spread from one chicken to another but a few, like Staphylococcus infection, and aspergillosis, are not.

Prevention by Sanitation

Sanitation is used to reduce the numbers of disease organisms which the chicken contacts to the level where they will no longer cause disease and to provide a clean, healthy environment. This can be done by cleaning and disinfecting, by adequate ventilation to reduce the number of organisms in the air and by reducing contact with other chickens by keeping them in cages.

Sanitation affects all levels of the birds' environment:

- *Building and Equipment:* Cages are intended to keep birds out of contact with their droppings, spilled feed and water, etc. that are a source of bacteria and bacterial growth. Fly control is important if this method is to be effective. On floor operations, an adequate amount of dry litter helps control bacterial growth.

  At cleanout, blow or spray dust from walls, ceiling, rafters, fans, etc. Wet down and remove litter, clean the building and equipment using detergent and water. Follow with a good disinfectant. A dirt flood cannot be disinfected. Litter should be piled or spread at least 100m from the poultry buildings (1 km is prefered).

- *Feed:* Between flocks, clean out feed in the troughs and bins both inside and outside the building and remove it from the building. It can be held to feed to chickens of a similar age or sent for reprocessing. Outside storage bins must be checked to make sure spoiled or moldy feed is not adhering to the inside and that the bin is empty. Feeders should be constructed and set so that feed is not spilled or billed out into the litter.
- *Water:* Open troughs are a source of contamination from the nasal and oral secretion and feces, etc. and must be cleaned regularly. Nipple drinkers are much cleaner but pressure must not be too high. Tremendous numbers of bacteria can build up in ponds and some wells, in water lines, and in drinking cups or troughs, so a periodic or constant water-sanitising programme and descaling may be necessary.
- *Air:* Clean, germ-free air is a very important part of a healthy environment. A good ventilation system to dilute and carry away dust and microorganisms and keep the pen dry is part of a sanitation programme.
- *The Caretaker:* Workers can carry infection to birds on their hands, clothes, boots, and equipment. They can spread germs from one group to another on the farm or bring germs in from outside the farm. If the same person is looking after more than one building, outer clothing and footwear should be changed between buildings and hands washed in disinfectant. If possible, separate workers should care for chicks, young birds and adults.

Good isolation requires shower-in and no contact with other chickens or other people who have chickens or work with chickens. Sanitation is a method of eliminating or reducing the number of disease causing organisms from contacting the birds. Although one germ may not cause disease and healthy birds may resist hundreds of germs, sanitation cannot be relied on to control serious contagious diseases. It must be used in conjunction with isolation and sometimes vaccination for problems such as Newcastle disease, avian influenza, infectious laryngotracheitis, fowl pox, etc. However, sanitation is important in maintaining the general health of a flock and is an important part of good flock management.

Prevention by Isolation

This method of disease control is simple. Stop the microorganisms that cause disease from contacting the chickens. When the microorganisms which cause a disease are eliminated from an area or country, the disease is said to be eradicated. Whether or not a disease can be prevented by isolation depends on:

(a) Where the microorganisms that cause the disease live. Some organisms (like the ones that cause necrotic enteritis, staphylococcosis, coccidiosis, *E. coli*, etc.) are widespread in the environment and it is difficult to prevent birds from contacting them. Some organisms are found primarily or only in chickens (sick birds or healthy carriers) and live outside the bird's body for only a short time (mycoplasma). Others may live for days, weeks, or months in the environment, depending on the organism and the conditions (moisture, temperature, etc.).

The way the disease organism is spread. Most disease germs are spread by direct contact, that is bird to bird. If the disease germs get on poultry fluff or small, air-borne particles, they can be carried by the wind, but this usually only over short distances. Organisms are frequently spread by mechanical transfer by carrier objects such as people, eggs, egg cases, trucks, feed, water, rats, dogs, insects, etc. Two or more separate objects may be involved in the spread of organisms from one flock to another (chickens to eggs, to egg cases, to people, back to chickens). Some animals, and certainly wild birds, can become infected with a disease and spread the germs. Insects are a necessary part of the spread of some diseases (leukocytozoon, plasmodium, etc.). Some disease organisms are spread through the egg from the adult to the chick (vertical spread). Some disease germs are spread venereally or by artificial insemination in semen (mycoplasma).

Diseases that are easier to control by isolation are those where the causative agent survives outside the body for only a short time. If the agent is egg-transmitted, the breeder flock must also be free of the disease. With some diseases a regular monitoring programme may be required to insure that the flock remains negative.

Isolation is not only the cheapest but the best way to control many contagious diseases and good isolation is equivalent to a quarantine.

Important Features of an Isolation Programme

Workers must be trained to carry out the sanitation and isolation procedures. They must also understand why the rules are set up and why disease prevention is important.

- Have only one age group on the farm (an all in, all out programme). Buildings over 100 meters apart can be treated as separate units if proper isolation and sanitation procedures are followed.
- Obtain chicks or replacements from a disease-free, adequately isolated, single source or raise replacements in a different area with separate caretakers.
- Have no neighbouring poultry buildings or free ranging chickens within 300 meters.
- Clean and sanitise buildings and equipment between crops. (wet down litter before removal to protect neighbouring poultry and do not store or spread litter near poultry buildings).
- Screen buildings against wild birds and keep out rats, cats, and dogs, and control insects. Remove dead birds from the pens at least twice a day and dispose of sick and dead birds at least 100m from the poultry buildings. Make sure dogs, cats and wild birds or animals cannot drag or carry dead chickens onto the farm.
- Limit the movement of workers from one building to another.
- Bring in only new or sterilised egg cases and flats.
- Make sure employees do not keep poultry or pet birds or come in contact with free-range chickens or their droppings and do not visit other poultry farms.
- Keep out visitors (particularly those who may visit other poultry farms) and provide boots and protective clothing for persons entering the poultry area.
- Disinfect necessary vehicles (feed trucks, etc.) and restrict them to the loading and unloading areas which should not be near the building entrance. Keep the driver in the truck or provide boots and coveralls.
- Make sure poultry service crews disinfect equipment, shower, and change clothing before entering the poultry area (except at cleanout).
- Shower and change clothing after taking chickens to market or meeting with poultry workers from other farms.

Prevention by Vaccination

Poultry have a good immune response to many diseases and to vaccination. They also pass immunity to offspring through the egg. Breeders require a special vaccination programme.

Marek's Disease (MD): There are several types of vaccine. All are live and given by injection. If vaccine is available, all chicks should receive MD vaccine immediately after hatching. Frequently two types of vaccine are given together. A second vaccination at a later date is not required. Chicks must be kept in a clean pen, away from other chickens for three weeks.

Newcastle Disease (ND): There is only one serotype of NDV so proper vaccination protects against all pathotypes of virus. Maternal antibody is provided for several days by immune hens. In countries where velogenic virus is endemic, chicks should be vaccinated before day 7 at day 21, 35 and 48 using more immunogenic (or virulent) vaccine (LaSota) for the 3rd and 4th vaccinations. Vaccinations should be repeated at 45 to 60 day intervals in places where ND is endemic. Properly administered live spray vaccine is best in adults. Eye drop administration is best for chicks. Either eye drop or spray can be used after 8 weeks. Vaccine in feed is available for village flocks. Killed vaccine may be given by injection.

Infectious Bronchitis (IB): There are many serotypes of IBV. Although there is some cross protection, new serotypes appear regularly that can evade the immunity of some vaccines. Commercial layers should be vaccinated as recommended locally. Broilers may also require protection. Live vaccines given in water or by spray are best in growing birds. Killed vaccines usually in oil may be used for adults. Some serotypes of IBV cause kidney disease.

Infectious Bursal Disease (IBD): There is one major serotype of IBDV but there are many pathotypes. As with ND, mild strains of vaccines will not protect against highly pathogenic strains of virus. Maternal antibody is provided for 7 to 14 days by immune hens. Where pathogenic strains of virus are present, moderate strength vaccine should be given at days 8, 14, and 28. If maternal antibody is not uniform, mild vaccine can be given at day 1. For maternal immunity killed vaccine in oil (often with ND & IB) may be given twice at 16 & 18 weeks and may be repeated at 35 to 45 weeks if necessary.

Avian Encephalomyelitis (AE): Epidemic tremors. Only breeders must be vaccinated but egg layers should be done as well to prevent a drop in egg production. Vaccination is done at about 12 weeks by water or wing web stab (live vaccine).

Viral Arthritis (VA): The arthritic form of reovirus infection is rare in many countries. Where the disease occurs broiler breeders should vaccinate twice to provide protection to chicks (live or killed vaccine).

Mycoplasma Gallisepticum (MG): Live MG vaccine is most effective if it is given to MG negative chickens before they become infected by the field strain. It provides less protection if given before 4 weeks. Other programmes are available for breeders.

Infectious Laryngotracheitis (ILT): ILT can be prevented by good isolation and sanitation procedures. The vaccine virus is live and may revert to pathogenic virus and is shed for life by vaccinated birds (as is field virus by survivors of outbreaks). In endemic areas breeders and egg type chickens should be eye drop vaccinated when moved to the laying house. Vaccinate 2 weeks before infection is likely to occur if growing birds are at risk. In some places broilers are vaccinated in the drinking water at 3 weeks.

Fowl pox: In endemic areas pox may be spread by mosquitos and cannot be prevented by isolation. Live pigeon pox (or other types) vaccine may be given by wing web stab at 8 to 12 weeks. Turkeys may also require protection.

Egg Drop Syndrome (EDS, Adenovirus Group III): In countries where EDS occurs layer and breeder flocks may be vaccinated at about 16 weeks with killed oil-adjuvant bacterin.

Adenovirus Group I (Inclusion body hepatitis, IBH): There are 12 or more serotypes of adenovirus Group I. Subtype 4 and 8 may cause disease without previous immunosuppression. Live vaccines may have to be prepared to prevent these infections.

Fowl Cholera: On farms or areas where fowl cholera occurs, birds may be vaccinated with killed or live vaccine. There are many strains of Pasteurella multocida and autogenous bacterins may be required for protection.

Infectious Coryza: As with fowl cholera. Live vaccine is not available.

Coccidiosis: Chickens or turkeys raised on litter or outdoors should be protected by medication until they develop immunity or by oral vaccination at day 1.

Turkeys may be vaccinated against Adenovirus Group II (which causes hemorrhagic enteritis (HE) with live vaccine and against erysipelas with bacterin.

Prevention by Medication

- Preventive medication Some diseases such as coccidiosis, necrotic enteritis and enterohepatitis can be prevented by medication. Preventive medication is most useful when protection is only required for a limited time as in broiler chickens or when immunity does not develop such as in necrotic enteritis.

Coccidiosis: Anticoccidial medication is very widely used in broilers and is generally used until the birds are ready for processing. Resistant strains of coccidia may develop and anticoccidials may have to be switched regularly. Immunity may or may not develop and anticoccidials that allow immunity are best.

Necrotic enteritis (NE) (Clostridium perfringens type A): Low level antibiotic in the feed is used to prevent NE and in Quail ulcerative enteritis (UE) (Clostridium colinum). The bacteria must be sensitive to the drugs being used and if predisposing factors such as coccidia infection are present they must also be corrected.

- *Therapeutic Medication:* Therapeutic medication can be considered preventive when it is used to control the spread of serious infectious diseases such as coryza or cholera. Treatment of coryza should not start until 1/3 of the flock is infected or the disease may recur following treatment.

  Some bacteria quickly develop resistance to medicine and different drugs may have to be used for control. Virus infections do not usually respond to medicine, although the medicine may prevent bacterial infection secondary to disease caused by virus.

  Medicines given by injection should not be given into the abdomen or leg. They can be given under the skin of the back or into the muscle of the breast.

Medicine given in the drinking water can be poured into the drinkers. In an automatic system they can be mixed in a large container and run into the system by gravity or a pump. Medicine can be added to a pressure system with a proportioner. Birds drink more water in hot weather. The level of medicine must match daily consumption and should be reduced in hot weather. It could be given for just 8 to 16 hours a day.

Medicine is often added to the feed at the feed plant. Most preventive medicine is used this way. In a disease outbreak medicine can be added to the water until medicated feed is available. Growing birds drink about 3 times as much water compared to the amount of feed they eat. Adults about 2 times as much (by weight).

Diagnosis of Poultry Disease

History. A good history will often provide clues that will help solve a problem. Get information on the type of bird, age, feed and water source and consumption rate, growth, production, morbidity and mortality, the description of the case, previous problems, vaccination programme, medicine being used, etc.

The problems may relate to management, environmental factors, and stress rather than to infection so examine the yard and housing conditions. Is the ventilation adequate? Are ammonia fumes a problem? Is it too hot or too cold? Is the litter wet or is it too dry and dusty? Is the pen too light or too dark? Are there sufficient hours of light for best production? Is the nest area darkened? Are the roosts too high? Do the birds appear comfortable? Chickens can talk and the sounds they make can indicate comfort, hunger, pain, panic, or disease.

Examination of Live Birds. Check the general appearance of the individual or group and try to determine which organ or system is involved in the illness. Note any signs or lesions that might point to a diagnosis, such as small size with poor feathering that suggests infectious stunting (malabsorption syndrome). If the birds show lameness or paralysis, is the lesion in the nervous system, bones, joints, muscles or skin? Some conditions, particularly those affecting locomotion, are easier to diagnose in live birds. Botulism which produces neck paralysis in chickens (leg and wing paralysis are more obvious in turkeys, ducks and pheasants) is an example.

Examine the skin of the head, body and legs for lice and mites, injury (particularly cannibalism), blood, mottling, swellings, anemia, cyanosis, or dermatitis. Listen for unusual breathing sounds (snicking, gurgling) and look for gasping or head-shaking that might indicate respiratory distress. Mouth-breathing (panting) is normal in chickens in hot weather. Exudate from nostrils and eyes and dirty feathers also suggest respiratory infection, or if just the eye, ammonia burn, ILTor eyeworm. Examine the droppings for evidence of diarrhea or other abnormalities. Take a blood sample for hematology or serology if indicated.

Necropsy. If a postmortem examination is to be carried out, birds that are representative of the problem in the flock must be selected. If birds have died, both sick and dead birds should be opened. Cull birds will not provide the answer. If the problem is a drop in production, try to find birds that look like they have recently stopped laying. It is important to do both an external and internal examination and to follow a specific routine to avoid missing important lesions.

- CRD - chronic respiratory disease
- IB(V) - infectious bronchitis (virus)
- ND - Newcastle disease
- AI - avian influenza
- CA(V) - chicken anemia (virus)
- EDS - egg drop syndrome
- IBH - inclusion body hepatitis
- ILT - infectious laryngotracheitis
- IBD - infectious bursal disease.

Live birds may be killed by cervical dislocation except when anemia or respiratory disease is suspected.

Examine the head, including the eyes, ears, nostrils, comb, wattles, mouth and beak.

Infected eyes and conjunctivitis are seen in CRD, sinusitis, Bordetella and Orthobacterium infection, infectious coryza and in ILT, IB and ND.

Keratoconjunctivitis with central corneal ulcers suggest ammonia burn. Swollen sinuses are seen in infectious coryza (Parahaemophilus infection) in chickens, sinusitis (mycoplasma), cryptosporidia and Bordetella infection in turkeys and pheasants. They may also be seen as the result of other infections such as influenza.

Chronic fowl cholera causes swollen wattles and sinuses in adult chickens. Fowl pox causes scabs on the comb, eyelid, and wattle, but must be differentiated from injury. Pox also occurs in turkey, pigeons, doves, canaries and other birds. If the bird has had its beak trimmed, check for proper healing, overgrowth of the lower beak or over-trimming (cut too short).

Injury on the head, neck or breast or back may indicate predators. Dermatitis and scabby or crusty lesions around the mouth and eyes suggest vitamin B deficiency.

Cut across the upper beak and back into the sinuses; then through the mandible and down the esophagus into the crop.

White plaques in the mouth, esophagus, or crop may be caused by capillaria worms, yeast infection (candidiasis, moniliasis), or possibly trichomoniasis or vitamin A deficiency, but most frequently by fowl pox (wet form). White plaques beside the tongue in the mouth are common in hens and turkey breeders and may be caused by mycotoxin or low humidity. Vomitoxin may produce similar lesions in young chickens and turkeys.

If the crop is enlarged and full, it may be an impacted or a sour crop (pendulous crop). The problem may be caused by excess water intake, defects in the crop itself, partial blockage of the proventriculus or gizzard or Marek's disease. Necrosis of the crop in wild birds is caused by Salmonella infection.

Examine the soft palate and larynx and cut down the trachea.

Wet fowl pox lesions are seen on the roof of the mouth, pharynx and larynx.

Granulation, congestion and mucus in the trachea are seen in CRD, IB, coryza, ILT, Orthobacterium, Bordetella, and *E. coli* infection, or with dust and ammonia fumes.

Hemorrhage and blood clots may occur in ILT and in ND or coryza and may cause severe gasping.

Gapes (gape worms) in pheasants, quail, and other birds is caused by Syngamus infection which also causes gasping. Cyanthastoma cause similar infection in water fowl and tracheal flukes may be found in waterfowl. Tracheal and air sac mites occur in cage birds.

Check the abdomen for lice and mites, and the vent for injury and prolapse of the oviduct.

Cut down between the abdomen and the legs and dislocate the hip joints. Peel the skin off the abdomen and breast. Tightly adhering skin and dark tissues indicate dehydration. Remove the breast carefully by cutting through the abdominal muscles, ribs and coracoids.

At this point, examine the thoracic and abdominal organs, paying particular attention to the air sacs, lungs, and liver. Carefully raise the gizzard and intestines and examine the abdominal air sacs, peritoneum, spleen and the ovary in laying birds.

The air sacs are cloudy in respiratory disease such as *E. coli* infection, Mycoplasma, Aspergillus, IB or early CRD in chickens. If there is also fibrin on the liver and in the pericardial sac, suspect *E. coli* or Salmonella infection. These lesions in turkeys and other birds might also be due to Chlamydia, or in ducks Anatipestifer infection.

Pneumonia in turkeys is caused by fowl cholera (Pasteurella multocida infection), and the lungs are quite solid. Aspergillosis, ND, AI, E. coli and Ornithobacterium infection can also cause pneumonia.

Peritonitis in layers is usually "egg peritonitis" caused by *E. coli* infection from the oviduct, although acute fowl cholera also causes peritonitis.

White crystals on the heart, liver and other tissues and organs are uric acid crystals (visceral gout) and are caused by hyperuricemia from urate nephrosis secondary to water deprivation, urolithiasis or other kidney disease. Large blood clots in the abdomen or hemorrhage and hematoma on the liver are the result of trauma or fatty liver hemorrhage syndrome in chickens, or ruptured aorta in turkeys.

Tumors in or on the organs may be Marek's disease, lymphoid leukosis, or other varieties of tumors. Multiple small tumors on the organs and peritoneum in adult hens are metastasis from a carcinoma of the oviduct. This may result in ascites. Ascites may also result from heart or liver disease or from ingestion of some toxic material. Right ventricular failure occurs most frequently in meat-type chickens after 4 weeks. It occurs secondary to pulmonary hypertension causing right ventricular hypertrophy and valvular insufficiency.

Focal white lesions on the organs may be tumors or they may be tuberculosis or coli-granuloma or, if just on the liver, blackhead. Turkeys, pheasants, and peacocks are all quite susceptible to blackhead; chickens are less susceptible. There are prominent cecal cores in blackhead (Histomoniasis) and occasionally in salmonella infection or coccidiosis. In chicks, turkeys and waterfowl, examine the lungs and air sac for yellow-white foci or plaques caused by brooder pneumonia (Aspergillosis). Gasping in young birds is a sign of tracheal or bronchial epithelial injury, or obstruction and can be caused by irritating fumes or by infection (often IB plus *E. coli* or Aspergillosis).

In young chicks, look for an infected navel and for yolk-sac infection or peritonitis (mushy chick, omphalitis) in which the abdomen is swollen, wet, and discoloured, and the yolk-sac is infected (due to *E. coli*, Salmonella, Staph., Proteus, etc.). Young birds also die because they don't start to eat (starve-outs) or drink (dehydration). Deaths occur mainly at 3, 4, and 5 days. This may be a management problem (chilling, feed and water not available, etc.) or the chicks/poults may be weak or defective when hatched. Broiler chickens that die suddenly from sudden death syndrome (dead in good condition) or from heat stroke, or suffocation (piling-up) have congested, edematous lungs, a full digestive tract and congested mottled breast muscle.

Examine the circulatory and immune systems, the heart, pericardial sac, blood, spleen, bursa, thymus, and lymphoid tissue of the thigh and intestine.

Birds that die from anemia are pale and the blood is watery. With CAV the thymus is small and bone marrow may be pale. Birds that bleed to death (pick-outs, ruptured fatty liver, acute cecal coccidiosis, ruptured aorta, hemorrhagic enteritis in turkeys, etc.) are also pale.

Sulfa poisoning produces anemia and widespread hemorrhage in the tissues. Chicken anemia virus (CAV) produces similar lesions and is the agent responsible for infectious anemia associated with IBH. Lead poisoning may also cause anemia. To identify anemia from parasites in blood cells (Plasmodium or leukocytozoon in ducks, turkeys and chickens), blood from a live, sick bird must be examined.

IBD (Gumboro disease) is caused by a virus that damages the bursa causing illness in 2-4 week old chickens or in younger chickens destroys part of the immune system, making birds more susceptible to other infections.

Swelling, congestion and hemorrhage with or without focal necrosis in the spleen, liver and lymphoid tissue suggest septicemia (fowl cholera, fowl typhoid, streptococcosis, or erysipelas) or viremia (ND and duck virus enteritis also affect the lymphoid tissue in the intestine). Marek's disease and lymphoid leukosis produce tumors in lymphoid tissue and organs except for the bursa which is mainly affected by lymphoid leukosis (occasional Marek's lesions may occur in the stroma of the bursa).

Skin leukosis is Marek's disease virus causing viral dermatitis in the feather follicles. At processing this can be confused with scabby hip or other causes of dermatitis. Marek's disease can also cause lymphoid neoplasia in the skin.

Cut through the proventriculus and remove the digestive tract and liver. Open the proventriculus, gizzard, and small and large intestines to the cloaca. Check the cloaca carefully for evidence of picking injury.

If hens are not properly beak-trimmed or are too fat, mortality from "pick-out" is common. The whole intestine may be picked out through the cloaca. Prolapse of the vagina (and cloaca) (blow-out) may occur from excess fat or straining, secondary to injury or inflammation. Injury is common in flocks that produce large eggs before they become mature.

Examine the digestive tract for lesions and the various kinds of enteritis (hemorrhagic, necrotic, ulcerative, etc.), parasites (tetrameres, roundworms, capillary worms, tapeworms, cecal worms, and coccidia), and gastrointestinal accidents. A large proventriculus in broilers is from lack of fibre in the diet resulting in poor development of the gizzard. A thickened proventriculus sometimes with ulcers or hemorrhage may be Marek's disease or infectious proventriculitis.

Green staining of the digestive tract is just bile and indicates that the bird is not eating. The liver and spleen may be small (if the bird is thin) and the gallbladder full. Salmonella pullorum (pullorum disease, bacillary white diarrhea) causes enteritis, diarrhea and death in chicks. It has been eradicated in many countries.

Check the ceca, intestine and liver for lesions of blackhead, TB, coccidiosis or tumor, the liver for other varieties of bacterial, viral or protozoal hepatitis, cholangiohepatitis, etc., and the pancreas for tumors. A large liver may be lymphoid leukosis or Marek's disease, or it may be caused by bacterial hepatitis (*E. coli*, campylobacter) or fowl typhoid (Salmonella gallinarum). Hepatitis with necrosis and hemorrhage that looks like bacterial (vibrionic) hepatitis may be immune damage to veins (vasculitis) from amyloid.

A large, yellow liver may be normal fat storage in a laying bird (estrogen stimulation) but layers die from fatty liver hemorrhage syndrome. This occurs when the liver ruptures because it has become fragile due to excess fat and free radical damage from fat. A yellow or hemorrhagic liver particularly with focal necrosis may be viral hepatitis (IBH) which is seen in broilers, pigeons, raptors, owls and psittacine (but IBH does not cause infectious anemia).

Examine the testes or ovary and open the oviduct. Shrinking ova indicate illness of 2-7 days' duration or one from which the bird may be recovering. Small, sac-like ova indicate that the bird has been out of lay for a week or more and may be in a molt.

Semi-solid (cooked) ova indicate bacterial infection (such as salmonella). An impacted oviduct may be secondary to vent-picking, egg-material left in the oviduct, or the bird may be egg-bound. Infection (Mycoplasma, IB, *E. coli*) can cause salpingitis as well.

A large or small fluid-filled cyst in the right abdomen beside the cloaca is the cystic remnant of the right oviduct (may be normal, but increased by IBV). A drop in production may be related to clinical or subclinical disease (EDS, IB, MG, AE, AI, ND, etc.) or management faults (lack of light, temperature change, lack of water) or nutritional problems, etc. Deformed shells suggest management faults, superovulation, EDS, IB and soft shells (higher than 1-2%) calcium or vitamin D3 deficiency.

A normal-appearing dead bird with an egg in the shell gland, or just laid has likely died from acute hypocalcemia. These birds often have fragile bones (particulary the femur), lack of medullary bone and rib infolding (osteoporosis, osteopenia, osteomalacia).

Hard (fibrotic) or swollen testes indicate bacterial infection (salmonella).

Examine the kidneys and ureters.

Swollen, pale or white spotted kidneys indicate hyperuricemia from urate nephrosis and may be due to lack of water or other kidney disease. Swollen. pale kidneys are also seen in IBH, IBD and fatty liver and kidney disease. Ureters plugged with hard stony material (urolithiasis) indicates a previous low phosphorus diet.

Swollen kidneys and nephritis may be due to IBV (nephrotrophic strain) or *E. coli* infection and usually causes death from dehydration. Newcastle's disease causes lympholytic foci in the kidney in pigeons.

Examine the skin, integument, muscles, bones and joints.

Emaciation, along with small organs, suggests malnutrition, stunting syndrome, beak injury (poor trimming), peck order (psychological) problems, chronic disease (coccidiosis), bumblefoot or other lameness, or chronic poisoning (lead, insecticide, etc.).

Muscular degeneration due to vitamin E-selenium deficiency can cause lameness, particularly in ducks. Ionophore toxicity causes muscle damage and paralysis in turkeys. Granulomas in the breast muscle are usually a vaccine reaction.

Sarcosporidial cysts produce small, white lesions in the muscle of water fowl.

Examine bones and joints for abnormality and deformity. Angular bone (valgus-varus) deformity of the intertarsal joint is caused by lateral or medial bending of the tibio-tarsal and metatarsal bones and is a common problem in meat-type poultry. It has a variety of possible causes (nutritional, rapid growth, management, etc.). Tibial dyschondroplasia causes backward bending or fracture of the top of the tibia. Slowing growth in young birds will help prevent leg deformity.

Other types of hock and stifle lameness and ruptured tendons are frequent in heavy roaster and turkeys and may be mainly due to injury as the result of heavy weight and fast growth, but some respond to added selenium or B vitamins.

Check for poor bone-breaking strength (osteoporosis) and, in young birds for rubbery bones, soft beaks, and beaded ribs which indicate calcium or Vit. D3 deficiency (rickets). Cage layer fatigue because of fragile bones may be due to phosphorus deficiency. Calcium and Vit. D3 deficiency also cause fragile bones (osteoporosis) in adults, but the most common cause is continuous high production.

Infectious stunting syndrome (malabsorption syndrome, fragile bones, osteoporosis) in young broilers is caused by intestinal damage from viral infection in young chicks. The chickens are small, poorly feathered and there is poorly digested food in the lower intestine. Osteomyelitis also causes fragile bones. Curly-toe paralysis in young birds may be riboflavin deficiency, but in older birds and turkeys may be genetic or due to lack of roosts. Cracked feet and foot dermatitis may be pantothenic acid or biotin deficiency, but scaly leg is caused by mites. Footpad dermatitis and hock lesion are often caused by poor litter conditions. Toe injury in young birds may be cannibalism or mechanical injury. Arthritis in the feet or hocks or other joints suggests infectious synovitis (Mycoplasma synoviae) or *E. coli* or Staph. infection often with osteomyelitis.

Infection in the wing joints in pigeons is usually due to salmonella.

In broilers, roaster and broiler breeders viral arthritis (reovirus infection) may cause lameness or ruptured tendons. Ruptured tendons are usually caused by rapid growth and large body size (heavy weight) and there is usually thickening of the tendon above the hock. If growing birds are lame, and there is no evidence of infection or rickets, cut into the proximal tibia and look for necrosis caused by osteomyelitis or dyschondroplasia and split the spine at T4 to look for spondylolisthesis (kinky-back), a plug of cartilage impinging on the cord or consider Marek's disease.

Necrotic dermatitis is caused either by staph. or clostridium infection and is associated with immunosuppression (usually by CAV or IBD). Scabby-hip is usually from overcrowding, poor litter conditions, poor feathering or scratches. Dermatitis or granulation in the neck may be a vaccine reaction, or contaminated vaccine.

Disturbances of the nervous system may cause incoordination, staggering, paralysis, walking backwards (with wings flapping for balance), tremors, stargasing, and other odd behaviour.

In case of lameness, paralysis, or incoordination, examine the sciatic nerves, spinal cord, and brain. Histologic examination will be required for diagnosis. ND may produce CNS disturbances in pigeons as well as nervous, respiratory, intestinal and reproductive lesions in chickens, pheasants, turkeys, and wild and cage birds of all ages.

Range paralysis is a form of Marek's disease affecting the peripheral and central nervous system causing lameness, incoordination and paralysis. Marek's can also cause botulism-like paralysis and transient paralysis.

Vitamin E deficiency (avian encephalomalacia) causes soft, dark areas of hemorrhage in the cerebellum which may be visible grossly. B vitamin deficiency (thiamine and other) may also cause nervous signs. Avian encephalomyelitis (AE) (epidemic tremor) affects birds up to 3-4 weeks old from non-immune parents.

Arsenilic acid and other feed additives and toxins (botulism, lead) may cause CNS disturbances, while others like the ionophores and coffee weed seeds (Cassia) cause nerve or muscle damage. Bacterial infection (pasteurella, pseudomonas, salmonella, staph. etc.) and fungi (aspergillus, etc.) also cause meningo-encephalitis, occasionally in outbreak proportions.

Hatching Eggs and Incubation

The Breeder Flock

Healthy chicks come from good breeders. Breeder nutrition, uniformity and the vaccination programme are critical.

Breeder Flock Health, Age and Nutrition

A healthy breeder flock, females with a good rate of lay mated to vigorous males, will usually produce highly fertile eggs that hatch well under good incubation practices. Hatching eggs from new breeder flocks (just starting to lay) do not hatch as well as eggs from those that have been laying for three to four months. Hatchability of eggs from old breeders nearing the end of their laying period also declines.

A hatching ration should be fed to breeders for about three weeks prior to saving eggs for incubation. The extra vitamins and minerals that it contains, as opposed to a laying ration, are essential for good hatchability. Sometimes, poor hatching results occur when a lower priced laying ration is used. The nutritional requirement of the breeder hen must be adequate for optimum hatching and good chick quality.

Clean Eggs are Important

Breeder house sanitation, nest box sanitation and hatching egg care are important to the production of healthy chicks. Nest material must be clean and dry and free of contamination. It should be cleaned daily (remove dirty material and droppings) and replenished or changed weekly. Para-formaldehyde pellets (20-25 g per nest, per month) will improve nest-box sanitation.

Breeder farms require a clean area where eggs can be cleaned, selected for quality and placed large end up on hatching trays. The temperature of this room should not be less than 21BC. If eggs are stored on the farm, temperature and humidity must be controlled.

Selection and Care of Hatching Eggs

Set only clean eggs. If eggs are soiled, it is preferable to clean them with sandpaper & egg brush. If hatching eggs are washed, the temperature of the water must be 43 to 44BC. The eggs should be spray washed with detergent, sanitised and fan-dried. Improperly washed eggs may rot or explode during the incubation period. Dirty eggs can be reduced if the eggs are gathered four or five times daily from nests supplied with fresh, clean nest material and if the litter in the breeding pens is kept dry so the hens feet are clean. Do not set floor eggs. Collect eggs with clean hands onto clean flats.

Select hatching eggs that are uniform in size (recommended minimum 52g for meat-type), shape and colour, with good sound shells. Do not set malformed, porous-shelled, doubled-yolked eggs or eggs with cracks. Shells that have a mottled appearance upon candling are not considered to have poor shell quality and can usually be set with good hatching results. Since smaller eggs hatch in less time than large eggs it is recommended that eggs be separated by size and that small eggs be put in the incubator 8-18 hours after the larger eggs are set.

The hatching percentage will be the highest if eggs are held at a temperature of 16 to 17BC for not more than one week before setting. Higher temperatures initiate embryo growth. Storage temperature should be reduced to 13BC if eggs are being held for two weeks or longer. Eggs may sweat when moved to warm, humid areas. This allows bacteria to penetrate the shell.

Turning eggs during the holding period is not beneficial. It has been shown that eggs held for more than 2 weeks hatched better when stored small end up (contrary to the accepted traditional large end-up postion). Relative humidity should be maintained at approximately 80% in the egg holding room. Higher humidity encourages mold growth. Prior to placing eggs in the incubator, they may be removed from the egg storage room and warmed to room temperature for approximately 6 hours.

The Incubator Room

In the selection of a successful incubator room, factors such as heating, humidity, ventilation, and sanitation should all be considered. Optimum results can be expected if the temperature can be maintained at between 24 and 27BC, with uniform humidity below the level that is required in the incubator.

Tropical climates (heat & humidity) make it difficult to maintain good incubator room conditions. Good ventilation and a constant supply of oxygen to remove excess carbon dioxide from the environment surrounding the incubating eggs is necessary for the developing embryo. High altitude reduces available oxygen. An incubator should not be placed near an outside wall or window in cold climates or in direct sunlight.

Incubators

Incubators are the most important equipment in the hatchery process. Many kinds of incubators are manufactured; however, the general principles of all modern machines for commercial hatchery production are the same. Incubator setting capacity ranges from approximately 14,000 to 100,000 eggs. During incubation, the hatching eggs are set vertically, with the large ends up in trays or flats in a setter and turned mechanically until about three days prior to hatching (setting period). The eggs are then transferred to a hatcher (hatching period) in a horizontal position and not turned during the hatching process. Both setters and hatchers have forced-draft air circulation, automatic temperature, humidity and cooling controls.

For small backyard poultry operators, there are small, still-air machines. Capacity of these incubators varies from 12 to 200 eggs. The eggs are set in a horizontal position and are usually turned manually. This type of incubator may be used for the entire incubation period for any kind of eggs.

The source of heat is a thermostatically controlled heating element or light bulb. Humidity is supplied in most cases by water in a pan below the eggs, and ventilation controlled by small air vents or holes. Most still-air machines have transparent plastic domes through which the incubation process can be observed.

Table: *Incubation Time for some Common Avian Species*

Species	Incubation time
Chicken	- 21 days
Turkey	- 28 days
Japanese quail	- 17 days
Guinea fowl	- 26 days
Pheasant, Partridge	- 24 days
Duck	- 28 days
Muscovy duck	- 35 days
Goose	- 28-32 days
Egyptian goose	- 35 days

The Principal Factors in Incubation

The five main factors affecting incubation, listed in order of importance are: temperature, humidity, ventilation, position, and turning of the eggs.

Some factors to consider before purchasing an incubator:

1. Any incubator will produce better results when operating at capacity. Egg capacity is a factor;
2. How long the incubator will last and cost of all replacement parts;
3. Service and availability of parts;
4. Accuracy of controls, to keep temperature and humidity fluctuations to a minimum;
5. The amount of labour involved in operating the machine and carrying out a thorough sanitation programme;
6. The guarantee.

Temperature

Hatching eggs may be warmed to a temperature of 25 to 30BC, prior to setting. The normal development of the embryo is dependent on the heat being held within a very narrow range in the incubator. In small still-air incubators, the temperature of the upper surface of the egg is higher than on the lower surface, while in large incubators, the air movement maintains the same temperature over the entire surface. For this reason, a still-air incubator must be operated at a higher temperature than a forced-draft incubator.

In small still-air incubators, a constant temperature of 39BC is considered satisfactory to produce good hatching results. The temperature may vary between 37.5 and 39.5BC without hurting the embryos as long as the temperature does not remain at either extreme. These readings should be taken with the bulb of the thermometer level with the upper surface of the eggs, but not in contact with the egg shell. A standing thermometer will give a more reliable reading than a hanging thermometer and the thermometer must be accurate. The temperature may rise one degree at hatching time without causing any reduction in hatch percentage.

In large incubators, the temperature, humidity, and speed of air movement are very closely dependent on each other, and since air speed varies in different incubators, it is impossible to state an exact operating temperature for all large machines, but generally it is around 37.5BC for a setter and 37BC for a hatcher.

Follow the manufacturer's instructions closely with regard to temperature and ensure that instructions are for the model in use. High temperatures even for a very short period of time during any part of the incubation period will cause more harm than low temperatures.

Numerous factors may contribute to high or low incubator temperatures.

- high or low room temperatures or floor temperature under the incubator affect the operating temperature of both large and small incubators. A large incubator placed near an outside wall may have one section operating at a lower or higher temperature than the other. These problems and those listed below will result in one group of eggs hatching earlier or later than the others, increased embryo mortality or cause leg deformity in chicks & poults.
- floor temperature variation because of drains causing cold or heat under the incubator.
- an incubator thermometer not reading correctly.
- the failure of automatic equipment such as thermostats, cooling coils or automatic dampers.
- improper air circulation leading in turn to a rise in temperature in part of the incubator, because of: a decrease in fan speed, usually caused by low voltage or a slipping fan belt; incorrect spacing of filled egg trays in a partially filled incubator.
- overloading an incubator for any one setting of eggs.

Humidity

During the incubation or setting period, eggs should lose 11 to 12% of their weight (another 3 to 4% in the hatcher, after day 18), due mainly to a loss of moisture. The amount of moisture (humidity) in the incubator controls the rate of evaporation from the egg.

The evaporation rate is also related to temperature, air speed, shell thickness, and size of eggs; the smaller the eggs, the greater percentage of moisture loss.

Too great a moisture loss from the egg in the early stage of incubation will cause the embryo to adhere to the shell, causing death. Insufficient evaporation may cause death from lack of oxygen because of a small air cell, since just prior to pipping the shell, the embryo pips into the air cell and starts to breath air.

The best guides to the correct amount of humidity in an incubator is the weight loss and the size and enlargement of the air cell during incubation, or the position at which the chick pips the shell. The degree of enlargement of the air cell should be determined by candling several eggs and estimating the averagee evaporation.

The amount of moisture in an incubator may be referred to as "relative humidity", which is a percentage of the moisture in the air at any given temperature. This can be measured by a wet-bulb thermometer.

A wet-bulb reading is based on air movement, and for this reason, a wet-bulb thermometer cannot be used to determine the amount of humidity in a still-air incubator.

The relative humidity for small incubators is usually stated as 60%. Other inexpensive but less accurate equipment is available to measure relative humidity. A level water pan on the floor of the incubator under the eggs during the entire incubation period will generally supply sufficient moisture to give good hatching results, provided the moisture content of the air in the environment is neither too high or too low.

Percent egg weight loss during the setting period can act as a guide to the correct humidity. This can be done by weighing some eggs at the start of incubation and the same eggs again on the 7th and 18th day, and calculating the loss that occurred (write the weight on the shell with a pencil). Chicken eggs should lose approximately 4.5% and 11.5% during the first 7 and the first 18 days of incubation respectively.

Eggshells thicker than 0.34 mm are too thick and humidity should be reduced to increase moisture loss. Below 0.31 mm is too thin. Normal would be 0.33 to 0.34 mm at the beginning of production. Shell thickness decreases with age of breeder hen with 0.31 mm normal at 60 weeks.

In large incubators, the temperature, air speed and humidity are intimately related, and the manufacturer's instructions should be followed closely.

Better hatching results may be obtained if the temperature is lowered and the humidity raised at hatching time (only if separate hatching compartments are available). Lowering the air temperature will provide an additional increase in relative humidity to keep shell membranes moist during the hatching process.

***Table:** Percent Relative Humidity At Different Wet-bulb Readings*

Wet-bulb Reading (BC)	***Relative Humidity (%) at Air Temperature of 37.2BC %***	***37.8BC %***
32.2	70	68
31.7	67	65
31.1	65	63
30.6	62	60
30.0	59	57
29.4	56	54
28.9	53	51
28.3	51	48
27.8	48	46
27.2	45	43
26.7	43	41

Incorrect humidity may be due to a number of factors, the more important of which are listed below:

1. High environmental humidity in tropical countries.
2. A wet-bulb thermometer reading incorrectly. Remove the wick to determine whether the thermometer is reading the same as the dry-bulb thermometer.
3. Dust and dirt on the wet-bulb wick. Change wicks often and use only distilled water in the reservoir.
4. In small incubators with humidity supplied by water pan evaporation, ensure that there is always an adequate amount of water in the pan, as the rate of evaporation is dependent on the humidity in the room. The lower the humidity in the room, the higher the evaporation and vice versa. During hatching fluff settles on the surface of the water producing a film that causes a reduction in the rate of water evaporation necessary for optimum hatchability. Clean the water pan daily and replace with clean, lukewarm water.
5. Ventilating an incubator to control temperature will lower humidity in a dry environment unless automatic humidifiers are being used in the incubator room.
6. If airflow in incubators is not uniform humidity may be different from bottom to top or side to side.

Ventilation

The free movement of oxygen, carbon dioxide and water vapour through the pores of the shell is important, since the developing embryo must be able to take in a constant supply of oxygen and release carbon dioxide and moisture. Oxygen content of 21% (present in air at sea level) and a carbon dioxide content not exceeding 0.5% in the air are considered optimum for good hatching results. Room temperature, room humidity, the number of eggs set, the period of incubation, and the air movement in the incubator all influence ventilation requirements. Ventilation problems are not the same in small incubators as they are in large incubators, where a large number of eggs are set in a very small space.

During the early part of the incubation period, ventilation in small incubators may be held to a minimum. However, during the hatching period additional ventilation must be supplied to reduce the carbon dioxide in the incubator. It is advisable not to increased ventilation until half of the hatch has been completed, since ventilating too soon will reduce the humidity. In large incubators, the manufacturer's directions should be followed, however, ventilating recommendations may not be applicable to every locality and every room condition.

If ventilation is used to control either temperature or humidity in the incubator, the control of the same factors in the incubator room are important. At a room temperature below 18 C, ventilating an incubator will reduce both temperature and humidity. In a room with high humidity, (tropical countries) the primary concern is to maintain the correct temperature.

The main ventilation consideration may be summarised as follows:

1. Ventilation is more important in large incubators than in small incubators.
2. The amount of ventilation required may be altered by atmospheric conditions.
3. Ventilation is very important in any incubator at hatching time. Insufficient ventilation may result in embryo or chick death.
4. Ventilation in excess of the recommended amount may be applied to reduce temperature or humidity.
5. The appearance of chicks panting in a hatcher at normal temperature is an indication of a rise in the carbon dioxide content of the hatcher air. Under such conditions chicks must breathe faster to obtain the required amount of oxygen and to eliminate the exceses carbon dioxide. If excessive panting occurs, increase the airflow in the hatcher.

Position and Turning of Eggs

In small incubators, the eggs are maintained in a horizontal position during the entire incubation period. In large incubators eggs should be placed in a vertical position, large end up, during the hatching period. In small incubators, the eggs are moved when turned, while in large incubators they remain in a stationary position on the incubator tray and the egg tray is turned through an angle of not less than 90 in opposite directions with each turning.

The objective is the same in both types of incubators; namely, to prevent the embryo from sticking to the shell membranes. Turning also ensures a complete contact of the embryonic membranes with the food material in the egg, especially in early stages of incubation.

In small incubators, the eggs should be turned at least four times daily. It is advisable to leave some space on the tray to allow for moving the eggs forward a 1/2 turn on one turn and back a 1/2 turn on the next, thus making sure that all the eggs move.

Eggs should not be turned in a complete circle, as this has a tendency to rupture the allantois sac with resultant embryonic mortality. Wash hands carefully before turning eggs to avoid bacterial contamination of the shell. In large incubators, the trays are usually turned hourly with all the egg trays moving at one time.

For good hatchability, eggs should be turned to a position at least 45 from vertical, then reversed in the opposite direction to a similar position. In the most recent models of incubators, eggs are turned through an arc of 150 and in a few models they are turned as far as 180. The introduction of these newer methods of turning eggs has been an important aspect in improving hatchability. Eggs should not be turned in either large or small incubators during the hatching period. The greatest benefit from turning eggs is during the first week in incubation.

Other Factors Affecting Incubation

Egg Selection

Poor quality hatching eggs do not hatch as well as eggs of good quality. The term "quality" refers to the condition outside the shell, the condition of the shell itself and that of the contents. Eggs with inferior characteristics, as discussed in "Selection and Care of Hatching Eggs," should not be set.

Sanitation

Eggs used for hatching should be clean and stored in clean containers in a sanitary egg holding room. Eggs contaminated with bacterial organisms usually do not hatch well and this poor quality is reflected in the chicks that do hatch.

Egg Handling

Rough handling of hatching eggs before they are set will increase the number of dead embryos, with mortality occurring between the 4th and 13th day of incubation. Also, jarring eggs during incubation may result in the rupture of the egg shell membrane and thereby lower hatchability. Large fluctuations in temperature and humidity during storage will have a major adverse affect on hatchability. Refer to "Selection and Care of Hatching Eggs" for proper egg storage procedure.

Toxicity

If the interior of an incubator is painted or varnished, or if the trays are varnished, the percentage of hatch will be reduced, possibly by as much as 25%. This adverse effect disappears in about 30 days, suggesting that the ill effect is eliminated by oxidation of the paint.

This problem may be overcome without any reduction in percentage of hatch if the incubator is fumigated with formaldehyde gas at the concentration recommended for proper hatchery fumigation. The gassing should be done as soon as the paint is dry and with the incubator operating at recommended temperature and humidity for incubating eggs.

Automatic Equipment

The addition of automatic equipment has eliminated many of the problems with incubators, however, such hazards may occur when automatic devices fail. During hatching all automatic devices, such as cooling coils, automatic dampers, tray turners, alarm bells, etc., should be checked at regular intervals. Electrical failure may require that automatic equipment be reset. After an incubator is purchased, it is important to know how every part operates.

Egg Candling

Candling chicken eggs on the 7th and 18th day of incubation, may be recommended for small poultry producers. Egg candling will detect infertiles and early dead germs. Therefore, problems within the hatching flock can be identified without waiting until the incubation period is completed.

Improper Fumigation

Closely follow instructions outlined in "Fumigation Procedure". Excessive and improper fumigation can result in high mortality in developing embryos.

Fumigation of Incubators

The killing of bacterial organisms by formaldehyde gas is based on the concentrations of the gas, exposure time, temperature, and humidity of the incubator. The chemicals potassium permanganate and formalin (which is 40% formaldehyde gas) have proven to be the most effective method of destroying bacterial organisms in the hatchery.

To accomplish the proper release of the gas, one and one-half parts (by volume) formalin is added to one part (by weight) of potassium permanganate. This will release the formaldehyde as gas or fumigant. When the reaction is complete, a dry, brown powder will be left. If the residue is wet, not enough permanganate was used; if the residue is purple, too much permanganate was added.

Concentration

The recommended concentration for effective fumigation is 53 mL of formalin added to 36 g of potassium permanaganate per cubic metre of space to be fumigated, or 1 1/2 mL of formalin added to 1 g of potassium permanganate per cubic foot of space to be fumigated.

Caution: Never add the permanagante to the formalin. Heat is generated when the two chemicals are combined, and care should be taken. Formaldehyde gas is generated quickly. Do not allow the fumes to get into the eyes. Personnel should use a respirator or wear a mask to avoid unnecessary exposure. Ventilate the incubator room to remove fumes that escape from the incubator.

Time

It is not recommended to fumigate setters with hatching eggs in them, but if such treatment becomes necessary, embryos between 24 and 96 hours of age should not be exposed to the above concentration of formaldehyde. Hatching compartments should be fumigated after the eggs are transferred from the setter to the hatcher, again after the hatch has been taken off and before the refuse has been removed from the trays, and finally after the hatcher has been thoroughly cleaned. Do not fumigate chicks with this concentration of formaldehyde gas. Small still-air incubators should be fumigated after the chicks have been removed and prior to discarding the refuse from the tray and again after the incubator has been thoroughly cleaned.

Fumigation Procedure

1. Make sure the temperature and humidity of the incubators are at normal operating conditions.
2. Measure the inside volume of the machine in cubic feet or cubic metres (length x width x height).
3. Close the ventilators, but leave the fans on.
4. Weigh the required amount of potassium permanganate into a wide enamelware or earthenware vessel large enough to accommodate the boiling and splattering action experienced when the formalin is added. Place the vessel and the permanganate in the area to be fumigated; then add the formalin.
5. Close the door immediately and leave closed for 20 minutes.
6. After 20 minutes, open the ventilators.
7. Open the doors of the machine for five minutes, leaving the fan on to allow more of the formaldehyde gas to escape, or neutralise it with a 25% solution of ammonium hydroxide equal to one-half the amount of formalin used. The hydroxide should be thrown directly on the floor of the machine and the doors closed. The formaldehyde gas will quickly be neutralised.

Continuous Fumigation of Hatcher

The greatest increase in bacterial organisms occurs during the hatching period. These can be reduced, but not completely eliminated, by slow release of formalin in the hatcher during the last 48 hours of the hatching period. For continuous fumigation to be effective, hatching eggs should also be fumigated at transfer time with the recommended concentration.

Place formalin in pan about 25 mm in depth, allowing 58 cm2 of pan for each cubic metre of hatcher space, or a pan about 1 in. deep, allowing 30 in2 of pan for approximately 1000 ft3 of hatcher space. Do not use permanganate. The pan should be placed in the open area of the hatcher in direct line with the airflow.

Place the pan of formalin in hatcher 48 hours prior to hatch completion. To overcome fluff deposited on the formalin, add more formalin about 24 hours before hatch is complete to increase evaporation.

If the fan stops (mechanical failure, etc.) and reduces the airflow, remove the pan of formalin immediately. If hatchers are only partially full, continuous fumigation is not recommended, as a reduction in airflow will create an excessive build up of formalin in the hatcher.

Effects of Fumigation

1. Properly carried out, fumigation should not affect hatchability.
2. Fumigation will only kill bacteria that are present on the surface of hatchery refuse. Fumigation will not kill bacteria inside unhatched or pipped eggs. It is important to dispose of hatchery refuse carefully to minimise hatchery contamination.
3. The hatchery room must be separate from the tray dumping room and from the chick processing area. Air flow and traffic must be controlled to prevent contamination of the chick processing and holding areas.
4. Efficient fumigation along with other sanitary measures should control navel infection (omphalitis).
5. Fumigation is not intended to replace a thorough cleaning programme.

Hatchery Design

Hatchery layout is important to a good sanitation programme. Arrange the hatchery so there is a one-way flow of material from the point where the hatching eggs are brought in to where the processed chicks go out. Hatcheries should be designed so the flow of eggs, chicks and personnel does not spread contamination from one room to the next. Doors, including one-way doors, help stop cross contamination between rooms. Positive air pressure prevents contamination through an open door. Workers should change outer clothing and wash hands before moving from one work station to another.

Summary of Some Important Factors

1. Feed breeder flock hatching ration that is well fortified with essential nutrients.
2. Use healthy breeding stock.
3. Provide good egg-holding facilities.
4. Avoid holding eggs in storage for more than one week.
5. Prewarm eggs for 6 to 8 hours at incubator room temperature.
6. Set clean, good quality eggs. Delay setting small eggs (those more than 10% less than average) for 8 to 16 hours.
7. Maintain correct incubation temperature, humidity and ventilation. Make sure air intake does not draw contaminated air into the incubator.

8. Turn hatching eggs frequently.
9. Maintain incubator room temperature between 21 and 24BC with good ventilation and relatively high humidity.
10. Fumigate regularly.
11. Clean vaccinating and beak trimming equipment. Newly hatched chicks may pick up contamination and infection in the hatchery from vaccinating and beak trimming equipment. This equipment requires a very rigid sanitation schedule.
12. Practice strict sanitation; cleanliness is very important for successful hatching operation. Make sure belts, equipment and workers hands used to move eggs or newly hatched chicks are kept clean.

Animal Welfare Problems in Chickens and Other Poultry Caused by Single Trait Genetic Selection for Meat and Egg Production

Chickens have several serious welfare problems that come from bad genetics and can be fixed only with good genetics The biggest problem in many intensively raised animals is pushing the animal's biology for more and more production. Breeders choose the most productive animals — the fastest growing, the heaviest, the best egg layers, and so on — and selectively breed just those animals. Bad things always happen when an animal is overselected for any single trait. Nature will give you a nasty surprise.

Bone breakage is a very serious problem in both caged and cage-free hens because laying hens have been overselected for egg production. Commercially bred hens put all their calcium and minerals into forming eggshells, and their own bones become depleted. Their bones are so weak that in cage-free systems a hen can break her leg just jumping off her perch. The only way to solve this problem is for the industry to accept the fact that birds with strong bones will produce slightly fewer eggs.

Laying hens have other problems, too, especially feather pecking and cannibalism. Feather pecking is what it sounds like: one hen pecks at another hen's feathers or pulls a feather out all the way. Severe feather pecking can lead to cannibalism, with the victim hen being wounded and then killed by the hen doing the pecking. Even though a feather-pecking hen can kill her victim, feather pecking probably isn't driven by the RAGE system. We know this because of studies mixing unfamiliar hens together. Aggression goes up, but feather pecking and cannibalism don't. They aren't the same thing.

Instead, feather pecking is probably displaced or redirected SEEKING behaviour. It's a kind of foraging or exploration of another bird instead of the ground. We know this because of research showing that chickens housed on litter do a lot less of it. They peck at the litter on the ground, not at each other's feathers. The more active the bird and the more foraging behaviour it does naturally, the more likely it is to develop severe feather pecking. Both feather pecking and cannibalism are affected by genetics.

Some modern broiler chickens have genetic problems related to growth. I was shocked to learn at a chicken-breeding seminar that the broiler chicken has been so overselected for rapid growth that its bone physiology is totally abnormal. In normal bone development, the body first "erects" a scaffolding or frame of cartilage and then fills in the frame with minerals that harden into bone. After the bone has hardened, the cartilage dies off through programmed cell death. In broiler chickens, something goes wrong with the cartilage, so the bones don't have support while they're hardening and end up misshapen. I liken it to building a new basement wall and taking down the plywood concrete forms before the concrete has fully hardened. In some of the worst cases, a chicken's feet are rotated almost 90 degrees and the legs are twisted. These chickens are genetically lame. Several studies have shown that lame broilers will choose feed laced with painkiller over their regular feed, and a study of lame turkeys showed that they started moving around a lot more once they were on painkillers. The industry has created chickens that have chronic pain in order to get birds that grow at the far outer limits of what is biologically possible. When an animal's biological system is pushed to the point where the physiology is totally pathological, I get disgusted.

The other problem is that modern broiler chickens have been bred to have stupendous appetites so they'll grow super-fast and reach market weight as soon as possible. The trouble is that the breeder chickens, the parents of the broilers, have the same stupendous appetites as their chicks. If you let a broiler breeder chicken eat everything she wants, she will become obese, her fertility will decline, and her life will be shortened.

These chickens have to be kept on a strict diet just to maintain normal weight. They act miserable, and many of them develop stereotypies. These birds have low welfare no matter what you do. If you let them eat all they want, they have bad welfare and if you don't let them eat all they want, they also have bad welfare. It's terrible. The industry is going to have to breed parent stock with smaller appetites. There's no other way to fix the problem.

Then there are other genetic problems that no one understands. One of the worst cases was the rapist roosters. I wrote about them in Animals in Translation. Fortunately, the broiler industry has made some genetic changes to correct these problems, although there's still a way to go. The rapist roosters violently attack hens and injure and even kill them. Before the 1990s there weren't any rapist roosters. They just suddenly appeared out of the blue. First it was just one strain of roosters that had become aggressive but within a couple of years almost all strains had developed the same behaviour. Nobody knows why.

The rapist roosters have two problems: They are hyperaggressive and they have stopped doing the courtship dance the hen needs to see before she will mate. They've lost the little piece of genetic code that makes them do the dance. When the hens don't see the courtship dance, they don't become sexually, which may make the roosters' aggression worse. An unreceptive hen would be a form of frustration because it is a restraint on the rooster's action. So the RAGE system would be activated to some degree.

When I wrote Animals in Translation it looked like the rapist roosters were a side effect of the industry's selective breeding programme to create chickens with bigger breasts for more white meat. But now researchers aren't sure what caused it, or whether the hyper-aggression and the bad courtship behaviour are the same problem or two different problems that happened at the same time. Industry breeding programmes are trade secrets. It's obvious the industry is selectively breeding for larger breast size because breast size is getting larger. But we don't know what other selective breeding programmes the industry might be using.

Ian Duncan has an interesting theory about what might have happened. Dr. Duncan points out that big-breasted male birds have trouble mating because their huge chests get in the way. Male turkeys have such big breasts now that they can't mate at all and the hens have to be artificially inseminated.

Dr. Duncan says that if the same thing is happening to male chickens, the broiler breeder industry may have misdiagnosed the problem. When broiler breeders see chickens with decreased fertility, they attribute the problem to low sex drive. It's possible that the breeders who created rapist roosters were actually trying to increase roosters' sex drive. If the breeders selected for higher libido they could have mistaken a little bit of aggressive behaviour towards the hen for higher sex drive and ended up breeding hyper-aggressive roosters that for some reason had also lost their courtship dance. We'll probably never know. Today the aggressive rooster problem has been greatly reduced although it hasn't been eradicated.

Better Breeding Strategies Using Group Genetic Selection

Most of the time breeders deal with genetic problems by culling chickens that have the problems and mating the ones that don't. Another interesting approach is group selection. The researcher Bill Muir at Purdue University has shown that you can reduce feather pecking genetically by using a technique called group selection. With group selection, instead of picking certain individuals, you pick certain family groups to breed. Dr. Muir has done this by raising several "sire family" groups — groups of chickens related to each other through their father — and then selectively breeding the group that has the highest egg productivity and the lowest amount of feather pecking and cannibalism.

Group selection has a couple of advantages over individual selection. First, the fact that you're working with groups instead of individuals means you know something about the birds' behaviour in a group. When breeders select high-productivity, individually housed laying chickens to breed, they don't know whether they are feather peckers or not because they've never lived with other birds. Individually housed laying hens can't express the behaviour.

Second, when you have a group of genetically related chickens living together you also see how living in a group affects their behaviour and productivity. When you choose which group to breed, you're not just choosing one genetic strain of chickens over another, or two desirable behavioural traits (good egg laying and low feather pecking) over undesirable traits. You're choosing one way of relating to the environment over another.

This is important because most behaviour is affected by what's going on in the environment. A hard-wired behaviour like the courtship dance is always the same no matter what's going on in the environment. It's like a computer subroutine; once you turn it on it just runs. But everything else is affected by the environment. When breeders use group selection instead of individual selection, they're factoring in the way the selected group relates to its social and physical environment.

Today, only a handful of companies provide all of the commercial layers and broilers around the world, which has greatly narrowed the gene pool. This has created a risky situation because genetically similar animals are vulnerable to the same diseases. Sure enough, when the Australians phased out their homegrown broilers and imported American birds, they ended up with more disease problems.

This is why it's important to preserve the old breeds of animals and poultry. Keeping the classic breeds alive is the only way to preserve genetic diversity and to save animals that have valuable genetic traits breeders may want to breed back into commercial lines in the future. The meat from some of the old breeds is more tender and better quality than meat from animals bred for rapid growth, and the chickens are hardier, too. They perform better in pasture-based or organic farms.

They are beautiful, unique animals that shouldn't be destroyed by commercial breeding. Fortunately, many of the older breeds of poultry and livestock are being raised by local farmers and sold in farmer's markets or to gourmet restaurants. If a serious disease ever kills commercial broilers or layers, the entire world will be thanking the small producers and hobbyists who have kept the old breeds of chickens from becoming extinct.

Reproductive Responses to Sel-Plex® Organic Selenium in Male and Female Broiler Breeders: Impact on Production Traits and Hatchability

Achieving success with broiler breeder management is like hitting a moving target. Modern broiler stocks have been reported to grow at 4.6 times the rate of a 1957 random-bred strain due to increased genetic potential (Havenstein et al., 2003a). The 6-fold improvement in carcass yield of 2001 stocks fed a 2001 diet compared to 1957 stocks fed a 1957 diet is 85-90% due to genetics, and 10-15% due to nutritional changes (Havenstein et al., 2003b). While broiler 42- day body weight is increasing each year, the 42-day target body weight for male and female broiler breeders has remained the same, or even decreased (Rustad and Robinson, 2002).

In 1979, Hubbard male and female breeders were approximately 50% of the 42-day broiler weight. In 2001, this percentage had decreased to 36.1 for males and 30.3 for females. In essence, the degree of feed restriction has continued to increase while there is increased competition for a reduced feed allocation.

Complications in managing this increased growth efficiency is further complicated by the development of 'yield' varieties, carrying increased amounts of breast muscle, often on a smaller carcass frame. As selection for broiler breeder egg production is not as heritable or profitable as selection for growth traits, there are continued increases in the growth potential while egg production is not emphasised. As a result, Whitehead (2000) indicates that geneticists continually compound the problem by breeding a bird that, if allowed to exist in its freely expressed adult state, is completely unfit for life.

Understanding the ovarian function of the chicken and its interaction with nutritional status, age, and strain is likely the most important issue affecting poultry breeding companies today. The process involves the conversion of genetic, environmental, and nutritional cues into a cascade of signals from the neuroendocrine system.

These signals must be integrated and responded to by the organs and tissues primarily involved in reproduction, which will in turn produce more signals for both local and distant activities. The resulting eggs produced are the net result of the bird's attempt to coordinate the demands its body and environment have placed on it. The ability of an embryo to survive the incubation process relies on a balance between hatchery management and breeder management. Specific feed ingredients, bird age, and flock management decisions can directly affect semen quality, the oviduct environment, and the egg environment. These factors combine to influence the potential of the egg to be fertile and ultimately to hatch.

Reproduction in the Broiler Breeder

The reproductive system of the laying hen is comprised of many organs. The list includes the hypothalamus, the anterior pituitary, the ovary, the oviduct, the liver and the skeletal system. Small follicle steroidogenesis, particularly estrogen production, is responsible for transforming a pullet into a hen. As plasma levels of estrogens increase, externally visible features include reddening and enlargement of the comb and wattles, a prenuptial feather molt (feather drop) and a widening of the pubic bones to permit egg passage. Internally, estrogen stimulates liver production of egg yolk lipids with a significant change in the colour and size of the liver. Finally, the oviduct enlarges and becomes a secretory organ for deposition of albumen.

The male focuses on quantity rather than quality when it comes to sperm production. The hen then must screen out unsuitable sperm in order to guarantee production of high quality chicks. Mature sperm spend the majority of their time in the oviduct. Following mating, the hen will store sperm in the highly specialised microenvironments of the sperm storage tubules (SST) are located in the vaginal region of the oviduct. Only about 1 to 2% of the originally inseminated sperm enter the SST (Bakst et al., 1994), where they exist in a near-dormant state. The survival of the sperm to the point of insemination depends on a combination of sperm quality and the ability of the hen to provide a safe environment for the sperm. The hormone messages being relayed between the ovary and the hypothalamic and pituitary control centers are altered by feed intake. Besides affecting follicle formation, and reproductive control, feeding level can alter the viability of the embryo through changes to the egg and to the early maturation process.

Excess nutrients are diverted into liver lipids, excess ovarian follicle development, and as abdominal fatpad (Etches, 1996). It can be a vicious cycle, with obesity continuing to worsen as the rate of egg production remains low and/or goes into early decline due to excess feed intake. The ovaries of growth-selected strains are particularly sensitive to overfeeding during the sexual maturation process. Body weight in broiler breeder hens has been reported to be negatively correlated with duration of fertility and fertile egg production. Ultimately, reduced chick production in overfed broiler breeders is the culmination of poor egg production combined with reduced fertility, hatch of fertile, and embryonic viability.

The female oviduct environment can be hostile to sperm despite their existence in the SST within the oviduct wall. Duration of fertility (measured by monitoring fertility in consecutive eggs) can be reduced under conditions of overfeeding. It is known that fewer sperm survive in some bird strains and when excess feed is used, but it is not clear how the surviving sperm are affected, and if the remaining sperm are of similar quality to the ones originally placed.

Factors Affecting Hatching Egg Quality

There are many factors that can affect the potential of the embryo to survive incubation and generate a quality chick. Some of these factors are out of our control, such as hen age, and others can be manipulated through management decisions, such as egg size and hatchery environment. It can be difficult to formulate diets to Optimise egg production, fertility, and hatchability as little is known about the nutritional requirements of the embryo.

Dietary vitamin levels are increased in the diet with the hope they will also be increased in the egg. Yet there can be adverse reactions with this type of approach, as some vitamins have antagonistic relationships with others. Furthermore, there can be stability issues for long-term storage, as well as for feed processing procedures. With current genetic stocks, if the chick hatches in a weakened state due to a vitamin or mineral deficiency, it is more likely to succumb to disease now than with previous stocks.

Growth-selected stocks have low immunoresponsiveness (Siegel et al., 1984) due to either inadvertent or intentional negative selection pressure combined with growth efficiency selection. The developing embryo is especially sensitive to vitamin deficiency, which will result in death, malformation or some other atypical response.

During embryo development, oxidative metabolism increases substantially over the incubation period and especially in the last few days before hatch. This normal respiration related to embryo growth results in the production of free radicals, which can cause tissue damage through lipid peroxidation, with polyunsaturated fatty acids being especially vulnerable. The chick has developed effective antioxidant pathways to prevent damage.

The primary defence mechanism is a group of three enzymes (superoxide dismutase, glutathione peroxidase, and catalase), which convert free radicals produced by cellular respiration into less harmful alcohols (Ursiny et al., 1997). A second level of defence are the natural antioxidants – vitamin E, carotenoids, ascorbic acid, and glutathione, which protect the developing chick (Surai, 1999). During the last week of incubation, fat-soluble antioxidants are moved into the liver and yolk sac membrane.

The major fat soluble antioxidant, vitamin E, moves from the yolk to the embryo tissue at this time. Ascorbic acid (vitamin C) is the major water-soluble antioxidant, and is produced in the yolk sac membrane before transport to tissues like the brain (Surai, 1999). This helps protect membrane lipids during the large metabolic effort of hatching. The third level of antioxidant defence is the generation of enzymes that rebuild damaged membranes.

3

Chick Production in Broiler Breeders

There are several important instances in the management of broiler breeders where "more" is not "better". In simple terms, "more feed" allotted to hens to the extreme degree of ad libitum feeding reduces settable egg production by as much as 40 eggs per hen.

Breeder hens use dietary energy differently than egg-type hens. Young broiler breeder hen respond to extra dietary energy by developing extra ovarian follicles. Several research projects have been carried out at the University of Alberta, that have shown that "more ovarian development" does not mean "better reproductive efficiency". Clearly, a high priority during the early laying period is to promote ovarian development at a rate that will result in the highest number of settable eggs. This paper will look at the process of sexual maturation of the breeder pullet to improve egg production, fertility and hatchability.

The critical time in broiler breeder management is the period from photostimulation (lighting) to peak production. This period is characterised by relatively fast weight gains, and the changes brought about by the hormones being produced by a newly active ovary. Photostimulation is generally considered the cue to initiate puberty, although the response to light is readily modified by level of feeding.

Light energy passes through the skull and illuminates the hypothalamus. When the hypothalamus receives a photostimulatory signal (long day length above a threshold of intensity), hormones involved in ovarian function are produced. One of the first responses you can see by looking at the ovary of the bird after lighting is that the very small ovarian follicles begin to increase in size. The small follicles produce large quantities of estrogen, the hormone that causes most of the reproductive transformation associated with puberty.

Firstly, estrogen influences the production of yolk precursors (building blocks of yolk production)in the liver of the bird. The liver visibly enlarges and becomes paler as its fat content increases for the production of egg yolk lipids.

Secondly, the oviduct increases in size so it is ready to receive ovulated follicles. This step appears to be a limiting one for heavy turkeys, as the ovary in these birds develops faster than the oviduct does, so there is a high incidence of follicles that are ovulated that are not picked up (internal ovulation) and hence, not formed into normal eggs.

Thirdly, estrogen influences changes to bone composition that allow calcium to be mobilised daily to make egg shells.

Finally, estrogen combines with male sex hormones (androgens) which results in changes to plumage, comb size and colour and sexual receptivity to males.

Shaver Starbro pullets were reared to 20 weeks of age to determine the influences of lighting programme and feed allocation from 20-25 weeks on ovarian development and production traits.

The experiment used two lighting programmes and two levels of feeding. A "slow photoperiod" treatment (SP) was gradually brought from 8L:16D with weekly increases in day length. A "fast photoperiod" treatment (FP) was increased from 8L:16D to 15L:9D in a single step at 20 weeks of age. The "slow feed" treatment group (SF) received very moderate increases in feed allocation from 20 to 25 wk of age. Increases in feed allocation of greater than 5 g, were divided up into two smaller increases per week.

The "fast feed" treatment group (FF) received a more generous feed increase between 20-25 weeks of age. It was not the original intent to decrease feed allocation to these birds; however, at 22 wk of age we discovered that the increases at 20 and 21 weeks of age had a greater effect on body weight that planned. The decision to decrease the feed allocation at 22 weeks of age was made to prevent the fast feed hens from becoming too heavy. It should be noted that the withdrawal of feed did not result in a decrease in body weight in subsequent weeks. From 25 to 64 weeks of age all of the birds were exposed to the same photoperiod (15L:9D) and a common feed allocation.

A group of birds were killed on the day following sexual maturity to examine development of the reproductive tract and carcass composition. Individual egg production records were kept and eggs were incubated at weekly intervals to assess treatment effects on chick production. Feed allocation was decreased after peak egg production to keep the birds at proper body weight targets (for age).

Body Weight: There were no significant differences in age at sexual maturity between any of the groups. Body weight and fat pad weights at sexual maturity were not different in terms of main effects. Within the group of slow photoperiod hens, the fast feed programme resulted in heavier body weights compared to the slow feed programme.

Ovary: Ovary morphology was highly influenced by both the photoperiod and feed allocation treatments. The slow photoperiod treatment had a greater ovary weight than did the fast photoperiod treatment. The fast feed treatment resulted in significantly heavier ovary weight and one more large follicle than the slow feed treatment. To our knowledge this is the first report of a relatively minor feeding difference affecting large follicle recruitment in commercial breeders.

Egg Production: All treatment groups had very high rates of total egg production when expressed in terms of main effects or the interactions of the main effects. There was a 10.9 egg advantage to use of the slow feed feeding programme compared to the fast feed programme. This fact is presumably related to improved control of the follicular recruitment process, as fast feed birds also had fewer large follicles at sexual maturity.

Fertility and Hatchability: Hatchability was reduced in the fast photoperiod programme compared to the slow photoperiod programme. Embryonic mortality was examined on day 7, 14, and 21 of incubation. The greatest embryonic mortality was found in the fast photoperiod treatment in the later stages of incubation.

Eggs from the fast fed hens had higher embryonic mortality than eggs from the slow fed hens. This indicates that developmental problems may be associated with the eggs from birds with excessive follicle development. These data suggest that minor differences in feed allocation have an influence on ovarian form and function. Proper management of feed allocation during the pre-lay phase is critical to Optimise the number of saleable chicks produced.

The above research data support the hypothesis that chick production in broiler breeder hens can be compromised by over-feeding early in lay. Future management programmes for breeder hens should focus on limiting excessive follicle production so that there is an orderly cooperation between the ovary and the oviduct in the production of hatching eggs.

Such a programme would minimise egg shell problems, reduce the incidence of multiple-yolked eggs and result in long egg-laying sequences. Long egg-laying sequences lead to high peak production and excellent persistency of lay. Further studies on limiting follicular growth by controlling energy intake and where nutrients are partitioned or distributed through the body and the role of light in sexual maturation are underway at the University of Alberta.

Broiler

A broiler is a type of chicken raised specifically for meat production. Modern commercial broilers, typically known as Cornish crosses or Cornish-Rocks are specially bred for large scale, efficient meat production and grow much faster than egg or traditional dual purpose breeds. They are noted for having very fast growth rates, a high feed conversion ratio, and low levels of activity. Broilers often reach a harvest weight of 4-5 pounds dressed in only five weeks.

They have white feathers and yellowish skin. This cross is also favourable for meat production because it lacks the typical "hair" which many breeds have that necessitates singeing after plucking. Both male and female broilers are slaughtered for their meat. In 2003, approximately 42 billion broilers were produced, 80% of which were produced by four companies: Aviagen, Cobb-Vantress, Hubbard Farms, and Hybro.

History

Before the development of modern commercial meat breeds (cows, chickens, etc.) broilers consisted mostly of young male chickens (cockerels) which were culled from farm flocks. The males were slaughtered for meat and the females (pullets) were kept for egg production. Compared to today, this made chicken meat scarce and expensive compared to eggs, and chicken was a luxury meat. The development of special broiler breeds decoupled the supply of broilers from the demand for eggs. This, along with advances in nutrition and incubation that allowed broilers to be raised year-round, allowed chicken to become a low-cost meat.

Broilers are often called "Rock-Cornish," referring to the adoption of a hybrid variety of chicken produced from a cross of male of a naturally double breasted Cornish strain and a female of a tall, large boned strain of white Plymouth Rocks. This first attempt at a hybrid meat breed was introduced in the 1930s and became dominant in the 1960s. The original cross was plagued by problems of low fertility, slow growth, and disease susceptibility, and modern broilers have gradually become very different from the Cornish x Rock hybrid.

Modern Variants

Access to a special diet of high protein feed delivered via an automated feeding system. This is combined with artificial lighting conditions to stimulate growth and thus the desired body weight is achieved in 4 - 8 weeks, depending on the approximate body weight required by the processing plant. After processing, the poultry is delivered as fresh or frozen chicken to the stores and supermarkets.

***Figure:** Five day old broiler strain Cornish-Rock chicks*

Because of their efficient meat conversion, broiler chickens are also popular in small family farms in rural communities, where a family will raise a small flock of broilers.

Broilers are sometimes reared on a grass range using a method called pastured poultry, as developed by Joel Salatin and promoted by the American Pastured Poultry Producers Association.

The term "broiler" is widely known in North America, Australia and England but not elsewhere in the English speaking world. The term "broiler chicken" is very widely used in Pakistan and India, as it was in the former German Democratic Republic and still nowadays in some eastern parts of Germany. The term is also used in Bangladesh, Indonesia, Sweden, Nigeria, Finland, Poland, Turkey and the Balkans.

Broiler Health Issues

Broiler chickens may develop several health issues as a result of selective breeding. Broiler chickens are bred to be very large to produce the most meat per animal. The large chickens cannot stand because their bodies grow too quickly for their legs. Therefore, they may become lame or suffer from broken legs. Broiler chickens are also

prone to heart attacks for the same reason, as the heart cannot support blood flow to the large body of the chicken. Another issue with selective breeding is the larger chickens have a more aggressive appetite. The broilers are feed restricted and this leads to behavioural issues in chronically hungry birds.

Broiler chickens may often get joint disorders because their legs cannot bear the heavy bodies. A Swedish study by SLU Skara (Swedish farming university) revealed that only 1/3 of studied broiler chickens that were about to be slaughtered were healthy. Additionally, it is very inactive and as a result is a poor forager, prone to predation, and is generally not suited to small free range homestead flocks.

If the litter in the pen is not properly managed to prevent birds from standing and resting in their feces, painful hock burns and foot ulcerations and blisters can occur. Pastured birds which are rotated frequently typically do not have these issues.

Leptin Receptor in the Chicken Ovary: Potential Involvement in Ovarian Dysfunction of Ad Libitum-fed Broiler Breeder Hens

The ovary of the mature hen contains a hierarchy of yellow yolky follicles and several thousand smaller follicles from which the large yolky follicles are recruited. The yellow follicles are arranged in a size hierarchy and are committed to ovulation. In each follicle, the granulosa cells are surrounded by theca tissue and are separated from it by a basement membrane. Each compartment of the largest yellow follicles (theca and granulosa cells) can be anatomically separated to follow the individual functions of these two ovarian compartments in follicle growth and differentiation.

At an early stage of follicular development the small ovarian follicles produce estrogens and androgens. As follicles begin to sequester yolk their production of estrogens from theca cells decreases to become very low at ovulation. As follicles are recruited into the yolky follicular hierarchy, estrogen and androgen production by theca cells diminishes and production of progesterone by granulosa cells increases. The largest F1 yellow follicle then attains the highest progesterone production at the time of ovulation.

In most domestic animals, reproductive function is considerably affected by nutrition. Various hormones, including growth hormone, insulin-like growth factors (IGFs) and insulin, have been proposed as potential mediators affecting reproductive function. However, the interactions between the reproductive endocrine axis and the metabolic

axis have not been clearly determined. Leptin represents also a good candidate for such reproductive-metabolic interactions. Leptin, the protein hormone synthesised and secreted mainly by adipose tissue, has primarily been shown to regulate food intake and energy expenditure. Recent studies have demonstrated that leptin may also be involved in the regulation of reproductive mechanisms in human and rat ovaries.

Exogenous leptin can rescue reproductive function in ob/ob leptin-deficient mice that are not only obese but also infertile. This leptin action is independent from weight loss since feed restriction in ob/ob female mice fails to restore fertility. Leptin can also advance the onset of puberty or at least reverse the delay caused by feed restriction in rodents. In chickens, leptin attenuates the negative effects of fasting on ovarian function.

Injections of leptin during fasting delays cessation of egg laying, attenuates regression of yellow hierarchical follicles, alters ovarian steroidogenesis and limits apoptosis. Leptin exerts its effect by binding to a receptor which belongs to the cytokine receptor super-family. The chicken leptin receptor has been cloned and sequenced. Its expression at the level of the ovary suggests that leptin might act directly on the ovary to regulate chicken reproductive function.

Standard broiler breeders have been submitted to high selection pressure on growth and feed efficiency. Male traits have been favoured resulting in poorer reproductive performances of the hens. As a result of such selection, broiler breeder hens are subject to metabolic disorders and reproductive dysfunction. Overfeeding during reproductive development is associated with the formation of excessive numbers of ovarian yellow follicles which can be arranged in multiple hierarchies, with increased production of unsettable eggs. Severe feed restriction during rearing reduces the production of yellow follicles, the incidence of double ovulation and considerably improves the laying rate. Up-regulation of yellow follicles has been related to excessive recruitment and rapid growth rate of follicles to maturity, especially under ad libitum feeding. However, the mechanisms that regulate these processes are still not fully explained.

This study investigated the potential involvement of leptin and its receptor in ovarian abnormalities observed in broiler breeder hens fed ad libitum. We aimed to determine the evolution of leptin receptor gene expression in both granulosa and theca cells from the four largest yellow follicles of 32 week-old hens during the laying period.

The effects of genotype and diet on plasma leptin levels and ovarian expression of the leptin receptor gene were also quantified.

For this purpose standard broiler breeder hens fed ad libitum or feed-restricted were compared to a French "Label" genotype and a dwarf "Experimental" line. Compared to fast growing standard broiler breeders, the French "Label" is a dwarf, slow-growing broiler genotype and the "Experimental" line is a dwarf genotype with a growth potential of progeny chicks close to that of the standard broiler chicks. The "Experimental" line is specifically selected for reproductive traits and viability at partial expenses of growth performances. The "Label" line tolerates ad libitum feeding of breeders and does not have reproductive problems under ad libitum feeding whereas the «Experimental» line can be fed ad libitum on a low energy diet during the growing period but presents more reproductive problems than the Label genotype but less than the standard genotype fed ad libitum.

Methods

Animals: Three lines of broiler breeder hens supplied by Hubbard primary breeder (Chateaubourg, France) were used in this experiment. The S line is a standard fast growing broiler, the French Label (L) line is a slow growing broiler breeder strain used for the quality market and the experimental (E) line is a broiler breeder strain bearing the "dw" dwarf gene. This strain has a decreased need for rationing. S and L hens were given the same regime in accordance with Hubbard nutritional recommendations (2724 kcal/kg) in fine meal form. During the growing period (0–20 weeks), half of the S hens were feed restricted (SR) on the same diet in order to match a reference body weight curve provided par Hubbard primary breeder, the other half (SA) and all the L hens were fed ad libitum.

The feed intake was equivalent to 37% of the SA group up to point of lay. A special diet was designed for the E hens, consisting of a series of finely ground meal diets with a lower energy content (2550 kcal/kg). The interest of the E group is mainly practical, it represents an actual alternative for severely restricted standard broiler breeder hens. Transition between the grower and breeder feed occurred at 20 weeks of age for the ad libitum-fed hens (L, SA and E). Transition occurred at the beginning of laying for the restricted hens that were then allowed ad libitum access to food.

At 32 weeks of age, blood samples were collected from 12 hens of each experimental group (SA, SR, E and L) and six hens were sacrificed by an overdose of pentobarbital (Sanofi-Sante Animale,

Libourne, France). The ovaries and liver were immediately removed. Granula and theca compartments from the first (F1), second (F2), third (F3) and fourth (F4) largest ovarian yellow follicles were dissected as previously described.

Since SA birds presented a greater average number of yellow follicles per ovary (9.36, 8,00, 7.42, and 6.33 yellow follicles/ovary for the SA, SR, E and L hens respectively) and a higher proportion of pairs of yellow ovarian follicles undergoing simultaneous development, follicles were assigned to the same follicular rank when the difference of weight between two follicles was of 0.4 g or less. In that case, one follicle per pair was collected and dissected. Tissues were immediately snap frozen in liquid nitrogen and stored at -80°C until used for total RNA extraction. This experiment was carried out with due regard to the legislation governing ethical treatment of animals, and investigators were certificated by the French government to carry out animal experiments.

RNA Extraction and Leptin Receptor RT-PCR

Total RNA was extracted from liver, granulosa and theca cells using RNA InstaPure (Eurogentec, Angers, France) according to the manufacturer's recommendations. After DNAse treatment using Ambion's DNA-free kit (Clinisciences, Montrouge, France), 2 μg of total RNA were reverse-transcribed (RT) in a final volume of 20 μl using RNAse H^- MMLV reverse transcriptase (Superscript II, Invitrogen, Cergy Pontoise, France) and random hexamer primers (Promega, Charbonnières, France). cDNA was then diluted to 1:8.

For normal PCR amplification, five microliters of the RT reaction were amplified for 35 cycles in a 50 μl reaction volume containing 2.5 units of Taq DNA Polymerase (Amersham Biosciences, Orsay, France), 2.5 mM $MgCl_2$, 0.2 mM dNTPs (Promega, Charbonnieres, France) and 0.2 μM of each forward and reverse primer. Leptin receptor forward (5'-GTC CAC GAG ATT CAT CCC AG-3') and reverse (5'-CCT GAG ATG CAG AGA TGC TC-3') primers were chosen according to the previously determined sequence of the chicken leptin receptor cDNA.

This pair of primers amplifies a 271 bp cDNA fragment located in the coding sequence of the extra-cellular domain. The amplification conditions were as follows: denaturation at 94°C for 30 sec, annealing at 60°C for 30 sec and primer extension at 72°C for 60 sec. After final extension at 72°C for 10 min, PCR products were resolved on 1.5% agarose gel containing ethidium bromide.

Real-Time RT-PCR

Real-time RT-PCR was performed as previously described. Briefly, forward leptin receptor primer 5'-GCATCTCTGCATCTCAGGAAAGA-3' and reverse leptin receptor primer 5'-GCAGGCTACAAA CTAACAAATCCA-3'(nucleotides 362 to 448 of the chicken leptin receptor cDNA sequence) were designed to be intron-spanning to avoid co-amplification of genomic DNA using Primer Express Software (Applied Biosystems, Courtaboeuf, France). A 20 μl master mix containing 12.5 μl SYBR Green PCR Master Mix, 1 μl forward primer (300 nM), 1 μl reverse primer (300 nM) and 5.5 μl water was prepared to perform real-time PCR (Applied Biosystems, Courtaboeuf, France). Five microliters of cDNA dilution was added to the PCR Master Mix to a final volume of 25 μl.

The following PCR protocol was used on the ABI Prism 7000 apparatus (Applied Biosystems, Courtaboeuf, France): initial denaturation (10 min at 95°C), followed by a two-step amplification programme (15 sec at 95°C, followed by 1 min at 60°C) repeated 40 times. Quantification was performed using ABI integrated software as previously described. 18S ribosomal RNA was chosen as the reference gene. The level of 18S RNA was determined using the Pre-developed TaqMan Ribosomal RNA control kit (Applied Biosystems, Courtaboeuf, France) according to the manufacturer's recommendations. The results were expressed as the leptin receptor mRNA/18S RNA ratio. Each PCR run included a no template control and replicates of control and unknown samples. Runs were performed in triplicate.

Plasma Lipid, Glucose and Hormone Analyses

Total cholesterol, phospholipid, and triglyceride plasma concentrations were calculated with "Cholesterol RTU", "Phospholipides Enzymatique PAP 150", and "Triglycerides Enzymatique PAP 150" kits (bioMérieux, Charbonnieres les Bains, France) according to the manufacturer's recommendations. Plasma glucose levels were measured by the glucose oxidase method using an automated analyser. Plasma insulin levels were determined by a radioimmunoassay with a guinea pig anti-porcine insulin antibody using chicken insulin as the standard. Plasma concentrations of leptin were determined by a multi-species leptin RIA kit (LINCO Research Inc, CliniSciences, Montrouge, France) according to the recommendations of the manufacturer.

Statistical Analysis

The results of plasma lipid, glucose and hormone analyses as well as leptin receptor mRNA expression in the liver were analysed by one-way ANOVA and means were compared by Student Newman Keuls multiple comparison test. The effects of the groups of birds (E, L SA and SR), follicular rank and possible interaction on the logarithm of leptin receptor mRNA levels were tested by two-way ANOVA using the General Linear Model (GLM) procedure of SAS (SAS Institute, 1999. SAS User' Guide, Version 8 ed. SAS Institute Inc., Cary, NC). An additional effect of the subject was introduced into the model in order to take into account the fact that measurements of leptin receptor mRNA expression for the different follicular ranks were performed on the same animal. Pairwise comparisons of means for each significant effect of the ANOVA were performed by Scheffe test with the least means squares statement of the GLM procedure. The level of significance was set at $P < 0.05$.

Results

Leptin Receptor MRNA Expression in Granulosa and Theca Cells: We demonstrated the expression of leptin receptor mRNA in granulosa and theca cells from the three genotypes fed ad libitum or restricted for the S line. The expression of leptin receptor mRNA for both ovarian cells was detected in each hierarchical yellow follicle studied (F1 to F4).

Evolution of Leptin Receptor MRNA Expression with Follicular Development

The evolution of expression of leptin receptor mRNA during follicle development was investigated in both granulosa and theca cells from F4 to F1 yellow follicles using real-time RT-PCR. Since leptin receptor mRNA levels did not follow a normal distribution (skewness of 5.48 and Kurtosis of 32.88), they were log transformed. The resulting distribution was closer to the normality with skewness of 0.87 and kurtosis of 0.69. Variance analysis was performed on transformed data.

In the E line, expression of the leptin receptor mRNA decreased between F4 and F1 yellow follicles. However the high variability of the expression of leptin receptor mRNA in F4 follicles prevented the decrease from reaching statistical significance. Compared to the L line, the level of expression of the leptin receptor in the granulosa cells was lower in the E group, especially for the F4 and F3 yellow follicles but statistical significance was reached only for the F4 follicles. In the S line fed ad libitum, expression of the leptin receptor in the granulosa was dramatically up-regulated. This up-regulation was clearly evident in F4 F3 and F1 follicles. Wide variability was also observed in F4 and F3 follicles. Feed restriction of the standard hens

(SR) induced a general decrease in the expression of leptin receptor mRNA. The overall level of expression of the leptin receptor mRNA measured in the SR line was similar to that observed in the L and E lines. Compared to the SA birds, feed restriction has restored the decrease in the expression of leptin receptor mRNA with follicle development.

Expression of leptin receptor mRNA in the theca cells remained stable during yellow follicle development, whatever the group of birds considered. Statistical analysis did not reveal any difference in leptin receptor mRNA expression between the 4 groups of birds.

Expression of Leptin Receptor MRNA in the Liver

In the liver, the expression of leptin receptor mRNA was up-regulated in S birds fed ad libitum. Feed restriction of S birds restored the level of expression of leptin receptor mRNA similar to that measured in the E and L birds.

Plasma Lipid and Glucose and Hormone Concentrations

At 32 weeks of age plasma leptin and insulin concentrations were found to be similar in the three genotypes. Food restriction of the standard hens did not alter plasma leptin, glucose or insulin levels. Triglyceride, cholesterol and phospholipid levels were also measured. Cholesterol and phospholipid levels were not affected by genotype or diet. On the other hand, triglyceride levels seemed to be affected by the genotype. Lower triglycerides levels were measured in Standard birds (SA, SR). However, statistical significance was reached only for the restricted SR birds.

Discussion

Several studies conducted on theca and granulosa cells have shown that leptin may have direct negative effects on ovarian steroidogenesis in various mammalian species. Leptin inhibits insulin-induced progesterone and 17β-estradiol production by isolated bovine granulosa cells and impairs the hormonally-stimulated in vitro release of 17β-estradiol by rat granulosa cells. In granulosa cells from fertile women, leptin inhibits FSH and IGF-I stimulated estradiol production Since leptin has a more potent inhibitory action of insulin-induced aromatase activity of granulosa cells from small than large follicles, it has been proposed that the numbers of leptin receptors in granulosa cells might decrease as follicles develop in order to make mature Graafian follicles less sensitive to the negative action of leptin. As shown in this study and in previous reports, the leptin receptor was expressed in the hen ovary in both granulosa and theca cells, suggesting a direct action of leptin at the level of the ovary.

It seemed that leptin might affect ovarian steroidogenesis in laying hens during fasting but the involvement of leptin on steroidogenesis during normal follicle development remained to be determined. In this study we demonstrated that the direct action of leptin on the ovary might be modified during follicle development since the level of expression of its receptor clearly decreased during maturation of yellow follicles. This decrease was particularly evident in slow growing broiler breeder hens from the "Label" genotype and from the feed-restricted standard line. Given that fast growing chickens (ad libitum-fed standard and Experimental broiler breeder hens) have the highest reproductive problems, genetic or nutritional control of the growth rate might regulate ovarian leptin receptor gene expression and improve reproductive function. Such evolution of receptor expression in the follicular hierarchy has previously been shown for the FSH receptor. FSH-stimulated steroidogenesis declined during follicle maturation and was associated with a decrease in FSH receptor numbers. Conversely, the expression of mRNA encoding the IGF-I receptor and the related efficacy of binding of IGF-I to granulosa cells increased as the follicle matured. Since leptin receptor gene expression was modified during follicle development, leptin might also be involved in regulation of the follicular hierarchy and onset of pre-ovulatory steroidogenesis, as has been proposed for gonadotrophins and growth factors including FSH and IGF-I.

Unlike mammals, progesterone in chickens is synthesised and secreted mainly by granulosa cells whereas theca cells generate estradiol. Progesterone produced by granulosa cells from mature follicles provides the positive feedback necessary to stimulate a pre-ovulatory surge of LH. IGF-I has been involved in the regulation of ovarian steroidogenesis in both mammals and birds.

IGF-I stimulates progesterone production from avian granulosa cells whereas it up-regulates estradiol from mammalian granulosa cells. Since leptin is considered to be an inhibitor of insulin and IGF-I action on steroidogenesis in mammals, leptin might have similar negative action in birds. Thus, the decrease in its receptor in the granulosa suggests that the inhibiting action of leptin would decrease during follicle development and consequently favours the stimulatory effect of gonadotrophins and IGF-I on follicular maturation.

This hypothesis is also consistent with the weaker steroidogenic response of granulosa cell culture of ad-libitum fed standard broiler breeder hens when stimulated by IGF-I compared to granulosa cell culture from feed-restricted birds. Moreover, Onagbesan et al (2004) demonstrated in an experiment similar to that performed in the present study and using the same genotypes that plasma progesterone levels were clearly affected in SA birds.

They demonstrated that plasma progesterone levels remained relatively stable between 25 and 37 weeks of age in the E, L and SA birds with a significant lower level in the SA birds (2.2 ± 0.62 ng/ml for SA birds compared to 3.9 ± 0.36 and 4.2 ± 0.54 for L and E birds respectively). In restricted standard birds, plasma progesterone levels dramatically increased and reached values (3.8 ± 0.26 ng/ml) closed to that measured in the E and L lines (Onagbesan et al., 2004, data from progesterone levels were personal communication from Dr Onagbesan, Catholic University of Leuven, Belgium).

The erratic pattern of oviposition in standard broiler breeder hens fed ad libitum has been previously demonstrated to be related to abnormal maturation of steroidogenesis, particularly in the two largest yellow follicles. Since F2 and F1 yellow follicles presented similar endocrine profiles, the pre-ovulatory surge of LH probably triggers ovulation of the two largest follicles. In this study we have shown that ad libitum feeding of broiler breeder hens dramatically up-regulated expression of the leptin receptor in the granulosa cells of yellow follicles and changed the evolution of expression of this receptor with follicle development. These results suggest a strong action of leptin on the ovaries of ad libitum fed birds.

Feed restriction reduced the level of expression of the leptin receptor and on the whole restored the evolution of expression of the receptor with follicle maturation. Since ad libitum feeding affects the hierarchical endocrine order of the follicles, as a potential inhibitor of hormonally induced avian steroidogenesis leptin represented a good candidate to explain the affects of follicular hierarchy. Up-regulation of the expression of the leptin receptor gene was also demonstrated in the liver. This up-regulation may be related to the control of lipogenesis. The liver plays a key role in lipid metabolism and lipogenesis in avian species and the standard broiler breeder hens were the fattest birds of this experiment.

Since expression of its receptor was dramatically up-regulated in SA hens, leptin probably played an important role in the increased number of large yellow follicles and abnormal follicle hierarchy. However the factors involved in regulation of the expression of the leptin receptor within the hen ovary remains to be determined. Among the plasma hormones and lipids analysed in this study only triglycerides were found to be different between strains, with a lower level in the restricted standard broiler breeder hens that were also the leanest birds. Down regulation of the expression of the leptin receptor by homologous and heterologous signals have previously been demonstrated in both mammals and chickens.

Leptin and insulin are able to down-regulate expression of the chicken leptin receptor in vitro. In the present study, plasma leptin and insulin levels were similar for each genotype and were not altered by feed restriction in the standard genotype. We have previously demonstrated that during the first 5 weeks of age, plasma leptin levels remained relatively stable in both broiler and layer chicken despite increased body weight.

However the absence of leptin levels differences may be related to the fact plasma leptin levels were measured at 32 weeks of age. SR birds were relaxed at the start of lay, switched to breeding feeding and allowed ad libitum access as the other groups of birds. We therefore suggested that leptin and insulin are probably not involved in the regulation of ovarian leptin receptor gene expression in ad libitum or feed-restricted standard broiler breeder hens.

Evidence of the regulation of expression of the leptin receptor gene in the granulosa related to follicle maturation and nutritional state strongly suggest that leptin played an important local and sequential role in the dysfunction of the follicular hierarchy observed in standard broiler breeder hens fed ad libitum. This study suggests that the level of expression of the leptin receptor regulates the action of its ligand at the level of the ovary. This provides an interesting perspective to understanding the physiological role of leptin in the ovary.

Chicken

The chicken (*Gallus gallus domesticus*) is a domesticated fowl, a subspecies of the Red Junglefowl. As one of the most common and widespread domestic animals, and with a population of more than 24 billion in 2003, there are more chickens in the world than any other species of bird. Humans keep chickens primarily as a source of food, consuming both their meat and their eggs.

The traditional poultry farming view of the domestication of the chicken is stated in *Encyclopaedia Britannica* (2007): "Humans first domesticated chickens of Indian origin for the purpose of cockfighting in Asia, Africa, and Europe. Very little formal attention was given to egg or meat production..." Recent genetic studies have pointed to multiple maternal origins in Southeast, East, and South Asia, but with the clade found in the Americas, Europe, the Middle East and Africa originating in the Indian subcontinent.

From India the domesticated fowl made its way to the Persianised kingdom of Lydia in western Asia Minor, and domestic fowl were imported to Greece by the fifth century BC. Fowl had been known in Egypt since the 18th Dynasty, with the "bird that lays every day" having come to Egypt from the land between Syria and Shinar, Babylonia, according to the annals of Tutmose III.

Terminology

In the UK, Ireland and Australia adult male chickens over the age of 12 months are primarily known as *cocks*, whereas in America and Canada they are more commonly called *roosters*. Males under a year old are *cockerels*. Castrated roosters are called *capons* (surgical and chemical castration are now illegal in some parts of the world). Females over a year old are known as *hens*, and younger females are *pullets*. In Australia and New Zealand (also sometimes in Britain), there is a generic term *chook* to describe all ages and both sexes. Babies are called *chicks*, and the meat is called *chicken*.

"Chicken" originally referred to chicks, not the species itself. The species as a whole was then called *domestic fowl*, or just *fowl*. This use of "chicken" survives in the phrase "Hen and Chickens", sometimes used as a British public house or theatre name, and to name groups of one large and many small rocks or islands in the sea. In the Deep South of the United States chickens are also referred to by the slang term *yardbird*.

General Biology and Habitat

Chickens are omnivores. In the wild, they often scratch at the soil to search for seeds, insects and even larger animals such as lizards or young mice.

Roosters can usually be differentiated from hens by their striking plumage of long flowing tails and shiny, pointed feathers on their necks (*hackles*) and backs (*saddle*) which are typically of brighter, bolder colours than those of females of the same species. However, in some breeds, such as the Sebright, the rooster has only slightly pointed neck feathers, the same colour as the hen's.

Figure: *The adult rooster can be distinguished from the hen by his larger comb*

The identification must be made by looking at the comb, or eventually from the development of spurs on the male's legs (in a few breeds and in certain hybrids the male and female chicks may be differentiated by colour). Adult chickens have a fleshy crest on their heads called a *comb or cockscomb*, and hanging flaps of skin either side under their beaks called *wattles*. Both the adult male and female have wattles and combs, but in most breeds these are more prominent in males. A *muff* or *beard* is a mutation found in several chicken breeds which causes extra feathering under the chicken's face, giving the appearance of a beard.

Domestic chickens are not capable of long distance flight, although lighter birds are generally capable of flying for short distances, such as over fences or into trees (where they would naturally roost). Chickens may occasionally fly briefly to explore their surroundings, but generally do so only to flee perceived danger.

Chickens are gregarious birds and live together in flocks. They have a communal approach to the incubation of eggs and raising of young. Individual chickens in a flock will dominate others, establishing a "pecking order", with dominant individuals having priority for food access and nesting locations. Removing hens or roosters from a flock causes a temporary disruption to this social order until a new pecking order is established. Adding hens—especially younger birds—to an existing flock can lead to violence and injury.

Hens will try to lay in nests that already contain eggs, and have been known to move eggs from neighbouring nests into their own. Some farmers use fake eggs made from plastic or stone (or golf balls) to encourage hens to lay in a particular location. The result of this behaviour is that a flock will use only a few preferred locations, rather than having a different nest for every bird. Hens can also be extremely stubborn about always laying in the same location. It is not unknown for two (or more) hens to try to share the same nest at the same time. If the nest is small, or one of the hens is particularly determined, this may result in chickens trying to lay on top of each other.

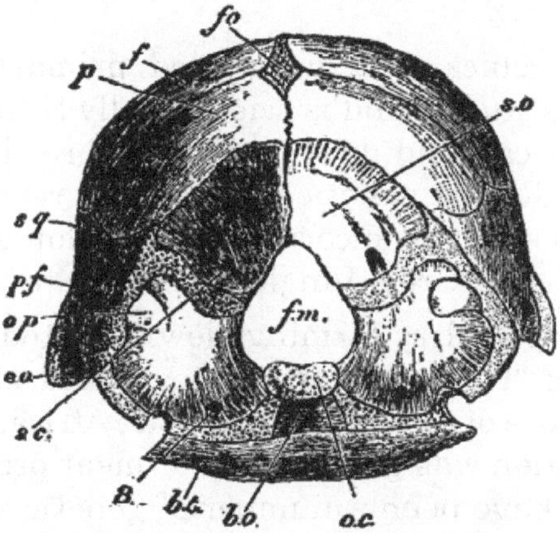

Figure: *Skull of a chicken three weeks old. Here the opisthotic bone appears in the occipital region, as in the adult Chelonian. bo = Basi-occipital, bt = Basi-temporal, eo = Opisthotic, f = Frontal, fm = Foramen magnum, fo = Fontanella, oc = Occipital condyle, op = Opisthotic, p = Parietal, pf = Post-frontal, sc = Sinus canal in supra-occipital, so = Supra-occpital, sq = Squamosal, 8 = Exit of vagus nerve.*

Roosters crowing (a loud and sometimes shrill call) is a territorial signal to other roosters. However, crowing may also result from sudden disturbances within their surroundings. Hens cluck loudly after laying an egg, and also to call their chicks. Chickens also give a low "warning call" when they think they see a predator approaching.

In 2006, scientists researching the ancestry of birds "turned on" a chicken recessive gene, *talpid2*, and found that the embryo jaws initiated formation of teeth, like those found in ancient bird fossils. John Fallon, the overseer of the project, stated that chickens have "...retained the ability to make teeth, under certain conditions...."

Food Sharing and Courting

When a rooster finds food, he may call other chickens to eat first. He does this by clucking in a high pitch as well as picking up and dropping the food. This behaviour may also be observed in mother hens to call their chicks and encourage them to eat.

To initiate courting, some roosters may dance in a circle around or near a hen ("a circle dance"), often lowering his wing which is closest to the hen. The dance triggers a response in the hen's brain, and when the hen responds to his "call", the rooster may mount the hen and proceed with the fertilization.

Breeding

Origins

The domestic chicken is descended primarily from the Red Junglefowl (*Gallus gallus*) and is scientifically classified as the same species. As such it can and does freely interbreed with populations of red jungle fowl. Recent genetic analysis has revealed that at least the gene for yellow skin was incorporated into domestic birds through hybridisation with the Grey Junglefowl (*G. sonneratii*).

The traditional poultry farming view is stated in *Encyclopaedia Britannica* (2007): "Humans first domesticated chickens of Indian origin for the purpose of cockfighting in Asia, Africa, and Europe. Very little formal attention was given to egg or meat production... " In the last decade there have been a number of genetic studies. According to one study, a single domestication event occurring in the region of modern Thailand created the modern chicken with minor transitions separating the modern breeds. However, that study was later found to be based on incomplete data, and recent studies point to multiple maternal origins, with the clade found in the Americas, Europe, Middle East, and Africa, originating from the Indian subcontinent, where a large number of unique haplotypes occur.

It has been claimed (based on paleoclimatic assumptions) that chickens were domesticated in Southern China in 6000 BC. However, according to a recent study, "it is not known whether these birds made much contribution to the modern domestic fowl. Chickens from the Harappan culture of the Indus Valley (2500-2100 BC), in what today is Pakistan, may have been the main source of diffusion throughout the world."

A northern road spread chicken to the Tarim basin of central Asia, modern day Iran. The chicken reached Europe (Romania, Turkey, Greece, Ukraine) about 3000 BC. Introduction into Western Europe came far later, about the 1st millennium BC. Phoenicians spread chickens along the Mediterranean coasts, to Iberia. Breeding increased under the Roman Empire, and was reduced in the Middle Ages. Middle East traces of chicken go back to a little earlier than 2000 BC, in Syria; chicken went southward only in the 1st millennium BC. The chicken reached Egypt for purposes of cock fighting about 1400 BC, and became widely bred only in Ptolemaic Egypt (about 300 BC). Little is known about the chicken's introduction into Africa. Three possible ways of introduction in about the early first millennium AD could have been through the Egyptian Nile Valley, the East Africa Roman-Greek or Indian trade, or from Carthage and the Berbers, across the Sahara.

The earliest known remains are from Mali, Nubia, East Coast, and South Africa and date back to the middle of the first millennium AD. Domestic chicken in the Americas before Western conquest is still an ongoing discussion, but blue-egged chicken, found only in the Americas and Asia, suggest an Asian origin for early American chickens.

A lack of data from Thailand, Russia, the Indian subcontinent, Southeast Asia and Sub-Saharan Africa makes it difficult to lay out a clear map of the spread of chickens in these areas; better description and genetic analysis of local breeds threatened by extinction may also help with research into this area.

Current

Under natural conditions, most birds lay only until a clutch is complete, and they will then incubate all the eggs. Many domestic hens will also do this–and are then said to "go broody". The broody hen will stop laying and instead will focus on the incubation of the eggs (a full clutch is usually about 12 eggs). She will "sit" or "set" on the nest, protesting or pecking in defence if disturbed or removed, and she will rarely leave the nest to eat, drink, or dust-bathe.

While brooding, the hen maintains the nest at a constant temperature and humidity, as well as turning the eggs regularly during the first part of the incubation. To stimulate broodiness, an owner may place many artificial eggs in the nest, or to stop it they may place the hen in an elevated cage with an open wire floor.

At the end of the incubation period (about 21 days), the eggs, if fertile, will hatch. Development of the egg starts only when incubation begins, so they all hatch within a day or two of each other, despite perhaps being laid over a period of two weeks or so. Before hatching, the hen can hear the chicks peeping inside the eggs, and will gently cluck to stimulate them to break out of their shells. The chick begins by "pipping"; pecking a breathing hole with its egg tooth towards the blunt end of the egg, usually on the upper side. It will then rest for some hours, absorbing the remaining egg yolk and withdrawing the blood supply from the membrane beneath the shell (used earlier for breathing through the shell). It then enlarges the hole, gradually turning round as it goes, and eventually severing the blunt end of the shell completely to make a lid. It crawls out of the remaining shell, and its wet down dries out in the warmth of the nest.

The hen will usually stay on the nest for about two days after the first egg hatches, and during this time the newly hatched chicks live off the egg yolk they absorb just before hatching. Any eggs not fertilized by a rooster will not hatch, and the hen eventually loses interest in these and leaves the nest.

After hatching, the hen fiercely guards the chicks, and will brood them when necessary to keep them warm, at first often returning to the nest at night. She leads them to food and water; she will call them to edible items, but seldom feeds them directly. She continues to care for them until they are several weeks old, when she will gradually lose interest and eventually start to lay again.

Modern egg-laying breeds rarely go broody, and those that do often stop part-way through the incubation. However, some "utility" (general purpose) breeds, such as the Cochin, Cornish and Silkie, do regularly go broody, and they make excellent mothers, not only for chicken eggs but also for those of other species—even those with much smaller or larger eggs and different incubation periods, such as quail, pheasants, turkeys or geese. Chicken eggs can also be hatched under a broody duck, with varied success.

Poultry Farming

More than 50 billion chickens are reared annually as a source of food, for both their meat and their eggs. Chickens farmed for meat are called broiler chickens, whilst those farmed for eggs are called egg-laying hens. In total, the UK alone consumes over 29 million eggs per day. Some hens can produce over 300 eggs per year. Chickens will naturally live for 6 or more years, but broiler chickens typically take less than six weeks to reach slaughter size. For laying hens, they are slaughtered after about 12 months, when the hens' productivity starts to decline, by which point they are normally infirm and have lost a significant amount of their feathers, and their life expectancy has been reduced from around 7 years to less than 2 years.

The vast majority of poultry are raised using intensive farming techniques. According to the Worldwatch Institute, 74 percent of the world's poultry meat, and 68 percent of eggs are produced this way. One alternative to intensive poultry farming is free range farming.

Friction between these two main methods has led to long term issues of ethical consumerism. Opponents of intensive farming argue that it harms the environment, creates human health risks and is inhumane. Advocates of intensive farming say that their highly efficient systems save land and food resources due to increased productivity, stating that the animals are looked after in state-of-the-art environmentally controlled facilities.

In part due to the conditions on intensive poultry farms and recent recalls of large quantities of eggs, there is a growing movement for small scale micro-flocks or 'backyard chickens'. This involves keeping small numbers of hens (usually no more than a dozen), in suburban or urban residential areas to control bugs, utilise chicken waste as fertilizer in small gardens, and of course for the high-quality eggs and meat that are produced.

Artificial Incubation

Incubation can successfully occur artificially in machines that provide the correct, controlled environment for the developing chick. The average incubation period for chickens is 21 days but may depend on the temperature and humidity in the incubator. Temperature regulation is the most critical factor for a successful hatch. Variations of more than 1 °F (1.8 °C) from the optimum temperature of 99.5 °F (37.5 °C) will reduce hatch rates.

Humidity is also important because the rate at which eggs lose water by evaporation depends on the ambient relative humidity. Evaporation can be assessed by candling, to view the size of the air sac, or by measuring weight loss. Relative humidity should be increased to around 70% in the last three days of incubation to keep the membrane around the hatching chick from drying out after the chick cracks the shell. Lower humidity is usual in the first 18 days to ensure adequate evaporation. The position of the eggs in the incubator can also influence hatch rates. For best results, eggs should be placed with the pointed ends down and turned regularly (at least three times per day) until one to three days before hatching. If the eggs aren't turned, the embryo inside may stick to the shell and may hatch with physical defects. Adequate ventilation is necessary to provide the embryo with oxygen. Older eggs require increased ventilation.

Many commercial incubators are industrial-sized with shelves holding tens of thousands of eggs at a time, with rotation of the eggs a fully automated process. Home incubators are boxes holding from half a dozen to 75 eggs; they are usually electrically powered, but in the past some were heated with an oil or paraffin lamp.

Chicken Eggs as Food

Chicken eggs are widely used in many types of dishes, both sweet and savory, including many baked goods. Eggs can be scrambled, fried, hard-boiled, soft-boiled, pickled, and poached. The albumen, or egg white, contains protein but little or no fat, and can be used in cooking separately from the yolk. Egg whites may be aerated or whipped to a light, fluffy consistency and are often used in desserts such as meringues and mousse. Ground egg shells are sometimes used as a food additive to deliver calcium. Some people prefer to just have a female, and raise it for the eggs.

Chickens as Food

The meat of the chicken, also called "chicken", is a type of poultry meat. Because of its relatively low cost, chicken is one of the most used meats in the world. Nearly all parts of the bird can be used for food, and the meat can be cooked in many different ways. Popular chicken dishes include roasted chicken, fried chicken, chicken soup, Buffalo wings, tandoori chicken, butter chicken, and chicken rice. Chicken is also a staple of many fast food restaurants.

Chickens as Pets

Chickens are sometimes kept as pets and can be tamed by hand feeding, but roosters can sometimes become aggressive and noisy, although aggression can be curbed with proper handling. Some have advised against keeping them around very young children. Certain breeds, however, such as silkies and many bantam varieties are generally docile and are often recommended as good pets around children with disabilities. Some people find chickens' behaviour entertaining and educational.

Chicken Diseases and Ailments

Chickens are susceptible to several parasites, including lice, mites, ticks, fleas, and intestinal worms, as well as other diseases. Despite the name, they are not affected by chickenpox, which is generally restricted to humans.

Some of the common diseases that affect chickens are shown below:

Name	*Common Name*	*Caused by*
Aspergillosis		fungi
Avian influenza	bird flu	virus
Histomoniasis	Blackhead disease	protozoal parasite
Botulism		toxin
Cage Layer Fatigue		mineral deficiencies, lack of exercise
Campylobacteriosis		tissue injury in the gut
Coccidiosis		parasites
Colds		virus
Crop Bound		improper feeding
Dermanyssus gallinae	Red mite	parasite
Egg bound		oversised egg
Erysipelas		bacteria
Fatty Liver Hemorrhagic Syndrome		high-energy food
Fowl Cholera		bacteria
Fowl pox		virus
Fowl Typhoid		bacteria
Gallid herpesvirus 1 or Infectious Laryngotracheitis		virus
Gapeworm	Syngamus trachea	worms
Infectious Bronchitis		virus
Infectious Bursal Disease	Gumboro	virus
Infectious Coryza		bacteria
Lymphoid leukosis		Avian leukosis virus
Marek's disease		virus
Moniliasis or Thrush	Yeast Infection fungi	
Mycoplasmas		bacteria-like organisms
Newcastle disease		virus
Necrotic Enteritis		bacteria
Omphalitis	Mushy chick disease	umbilical cord stump
Peritonitis		Infection in abdomen from egg yolk
Prolapse		
Psittacosis		bacteria
Pullorum	Salmonella	bacteria
Scaly leg		parasites
Squamous cell carcinoma		cancer
Tibial dyschondroplasia	speed growing	
Toxoplasmosis		protozoal parasite
Ulcerative Enteritis		bacteria
Ulcerative pododermatitis	Bumblefoot	bacteria

Chickens in Religion and Mythology

In Indonesia the chicken has great significance during the Hindu cremation ceremony. A chicken is considered a channel for evil spirits which may be present during the ceremony. A chicken is tethered by the leg and kept present at the ceremony for its duration to ensure that any evil spirits present during the ceremony go into the chicken and not the family members present. The chicken is then taken home and returns to its normal life.

In ancient Greece, the chicken was not normally used for sacrifices, perhaps because it was still considered an exotic animal. Because of its valour, the cock is found as an attribute of Ares, Heracles, and Athena. The alleged last words of Socrates as he died from hemlock poisoning, as recounted by Plato, were "Crito, I owe a cock to Asclepius; will you remember to pay the debt?", signifying that death was a cure for the illness of life.

The Greeks believed that even lions were afraid of cocks. Several of Aesop's Fables reference this belief.

In the New Testament, Jesus prophesied the betrayal by Peter: "Jesus answered, 'I tell you, Peter, before the rooster crows today, you will deny three times that you know me.'" (Luke 22:34) Thus it happened (Luke 22:61), and Peter cried bitterly. This made the cock a symbol for both vigilance and betrayal.

Earlier, Jesus compares himself to a mother hen when talking about Jerusalem: "O Jerusalem, Jerusalem, you who kill the prophets and stone those sent to you, how often I have longed to gather your children together, as a hen gathers her chicks under her wings, but you were not willing." (Matthew 23:37; also Luke 13:34).

In many Central European folk tales, the devil is believed to flee at the first crowing of a cock.

In traditional Jewish practice, a kosher animal is swung around the head and then slaughtered on the afternoon before Yom Kippur, the Day of Atonement, in a ritual called kapparos. A chicken or fish is typically used because it is commonly available (and small enough to hold). The sacrifice of the animal is to receive atonement, for the animal symbolically takes on all the person's sins in kapparos. The meat is then donated to the poor. A woman brings a hen for the ceremony, while a man brings a rooster. Although not actually a sacrifice in the biblical sense, the death of the animal reminds the penitent sinner that his or her life is in God's hands.

The Talmud speaks of learning "courtesy towards one's mate" from the rooster (Eruvin 100b). This might refer to the fact that when a rooster finds something good to eat, he calls his hens to eat first.

The chicken is one of the Zodiac symbols of the Chinese calendar. Also in Chinese religion, a cooked chicken as a religious offering is usually limited to ancestor veneration and worship of village deities. Vegetarian deities such as the Buddha are not one of the recipients of such offerings. Under some observations, an offering of chicken is presented with "serious" prayer (while roasted pork is offered during a joyous celebration). In Confucian Chinese Weddings, a chicken can be used as a substitute for one who is seriously ill or not available (e.g. sudden death) to attend the ceremony. A red silk scarf is placed on the chicken's head and a close relative of the absent bride/groom holds the chicken so the ceremony may proceed. However, this practice is rare today.

A cockatrice was supposed to have been born from an egg laid by a rooster, as well as killed by a Rooster's call.

Chickens in History

An early domestication of chickens in Southeast Asia is probable, since the word for domestic chicken (**manuk*) is part of the reconstructed Proto-Austronesian language. Chickens, together with dogs and pigs, were the domestic animals of the Lapita culture, the first Neolithic culture of Oceania.

The first pictures of chickens in Europe are found on Corinthian pottery of the 7th century BC. The poet Cratinus (mid-5th century BC, according to the later Greek author Athenaeus) calls the chicken "the Persian alarm". In Aristophanes's comedy *The Birds* (414 BC) a chicken is called "the Median bird", which points to an introduction from the East. Pictures of chickens are found on Greek red figure and black-figure pottery.

In ancient Greece, chickens were still rare and were a rather prestigious food for symposia. Delos seems to have been a centre of chicken breeding.

The Romans used chickens for oracles, both when flying ("ex avibus", Augury) and when feeding ("auspicium ex tripudiis", Alectryomancy). The hen ("gallina") gave a favourable omen ("auspicium ratum"), when appearing from the left (Cic.,de Div. ii.26), like the crow and the owl.

For the oracle "ex tripudiis" according to Cicero (Cic. de Div. ii.34), any bird could be used, but normally only chickens ("pulli") were consulted. The chickens were cared for by the pullarius, who opened their cage and fed them pulses or a special kind of soft cake when an augury was needed. If the chickens stayed in their cage, made noises ("occinerent"), beat their wings or flew away, the omen was bad; if they ate greedily, the omen was good.

In 249 BC, the Roman general Publius Claudius Pulcher had his chickens thrown overboard when they refused to feed before the battle of Drepana, saying "If they won't eat, perhaps they will drink." He promptly lost the battle against the Carthaginians and 93 Roman ships were sunk. Back in Rome, he was tried for impiety and heavily fined.

In 161 BC, a law was passed in Rome that forbade the consumption of fattened chickens. It was renewed a number of times, but does not seem to have been successful. Fattening chickens with bread soaked in milk was thought to give especially delicious results. The Roman gourmet Apicius offers 17 recipes for chicken, mainly boiled chicken with a sauce. All parts of the animal are used: the recipes include the stomach, liver, testicles and even the pygostyle (the fatty "tail" of the chicken where the tail feathers attach).

The Roman author Columella gives advice on chicken breeding in his eighth book of his treatise on agriculture. He identifies Tanagrian, Rhodic, Chalkidic and Median (commonly misidentified as Melian) breeds, which have an impressive appearance, a quarrelsome nature and were used for cockfighting by the Greeks. For farming, native (Roman) chickens are to be preferred, or a cross between native hens and Greek cocks. Dwarf chickens are nice to watch because of their size but have no other advantages.

Per Columella, the ideal flock consists of 200 birds, which can be supervised by one person if someone is watching for stray animals. White chickens should be avoided as they are not very fertile and are easily caught by eagles or goshawks. One cock should be kept for five hens. In the case of Rhodian and Median cocks that are very heavy and therefore not much inclined to sex, only three hens are kept per cock. The hens of heavy fowls are not much inclined to brood; therefore their eggs are best hatched by normal hens. A hen can hatch no more than 15-23 eggs, depending on the time of year, and supervise no more than 30 hatchlings. Eggs that are long and pointed give more male, rounded eggs mainly female hatchlings.

Per Columella, chicken coops should face southeast and lie adjacent to the kitchen, as smoke is beneficial for the animals. Coops should consist of three rooms and possess a hearth. Dry dust or ash should be provided for dust-baths.

According to Columella, chicken should be fed on barley groats, small chick-peas, millet and wheat bran, if they are cheap. Wheat itself should be avoided as it is harmful to the birds. Boiled ryegrass (*Lollium* sp.) and the leaves and seeds of alfalfa (*Medicago sativa* L.) can be used as well. Grape marc can be used, but only when the hens stop laying eggs, that is, about the middle of November; otherwise eggs are small and few. When feeding grape marc, it should be supplemented with some bran. Hens start to lay eggs after the winter solstice, in warm places around the first of January, in colder areas in the middle of February. Parboiled barley increases their fertility; this should be mixed with alfalfa leaves and seeds, or vetches or millet if alfalfa is not at hand. Free-ranging chickens should receive two cups of barley daily. Columella advises farmers to slaughter hens that are older than three years, because they no longer produce sufficient eggs.

Capons were produced by burning out their spurs with a hot iron. The wound was treated with potter's chalk.

For the use of poultry and eggs in the kitchens of ancient Rome see Roman eating and drinking.

Chickens were spread by Polynesian seafarers and reached Easter Island in the 12th century AD, where they were the only domestic animal, with the possible exception of the Polynesian Rat (*Rattus exulans*). They were housed in extremely solid chicken coops built from stone.

Chickens in South America

An unusual variety of chicken that has its origins in South America is the araucana, bred in southern Chile by Mapuche people. Araucanas, some of which are tailless and some of which have tufts of feathers around their ears, lay blue-green eggs. It has long been suggested that they predate the arrival of European chickens brought by the Spanish and are evidence of pre-Columbian trans-Pacific contacts between Asian or Pacific Oceanic peoples, particularly the Polynesians and South America.

In 2007, an international team of researchers reported the results of analysis of chicken bones found on the Arauco Peninsula in south central Chile. Radiocarbon dating suggested that the chickens were

Pre-Columbian, and DNA analysis showed that they were related to pre-historic populations of chickens in Polynesia. These results appeared to confirm that the chickens came from Polynesia and that there were transpacific contacts between Polynesia and South America before Columbus's arrival in the Americas.

However, a later report looking at the same specimens concluded:

> *A published, apparently pre-Columbian, Chilean specimen and six pre-European Polynesian specimens also cluster with the same European/Indian subcontinental/Southeast Asian sequences, providing no support for a Polynesian introduction of chickens to South America. In contrast, sequences from two archaeological sites on Easter Island group with an uncommon haplogroup from Indonesia, Japan, and China and may represent a genetic signature of an early Polynesian dispersal. Modelling of the potential marine carbon contribution to the Chilean archaeological specimen casts further doubt on claims for pre-Columbian chickens, and definitive proof will require further analyses of ancient DNA sequences and radiocarbon and stable isotope data from archaeological excavations within both Chile and Polynesia.*

Drinking and Feeding Systems

Feeding Systems

There are several different systems available for delivery and distribution of feed to broilers. Since feed constitutes the major share of total production cost, wastage should be an important consideration in the choice of system.

There are three major systems available:

- Automatic pan feeders: 1 pan per 65 birds; 33 cm pan diameter.
- Chain feeders: 2.5 cm per bird; 80 birds per metre of track.
- Round, hanging tube feeders: 65 birds per tube; 38 cm diameter base.

Automatic pan feeding systems have become the industry standard due to advantages of low feed wastage, ease of height adjustment, preservation of pellet quality, and reliability. As a number of different pan feeder designs are available, feeder heights should be set according to manufacturers' recommendations. Distance between the feeder lines should be not more than 2.5 metres. This ensures that all birds have adequate access to feed. Level of feed within the feeder should be adjusted to a height that minimises wastage. If possible, the feed supply system should be allowed to empty at least once a day. This eliminates the presence of stale food and therefore reduces the risk of contamination and the growth of micro-organisms.

Drinking Systems

It is essential that fresh water is available to the broiler flock at all times and that it is free of contamination. The drinking systems chosen must be capable of delivering the water efficiently to all birds with the minimum of spillage. To ensure that the flock is receiving sufficient water, each day, the ratio of water to feed consumed should be monitored.

When the ratio of water volume (ml or l) to feed weight (g or kg) remains close to 1.8:1 (1.6:1 for nipple drinkers), only then can it be assumed that the birds are consuming sufficient water. Birds will drink more water at high ambient temperatures. Water requirement increases by approximately 6.5% per degree as temperature exceeds 21°C. Water consumption will vary with feed consumption.

Nipple Drinkers

Nipple systems provide water with lower levels of bacterial contamination than conventional open systems. They have become the standard in modern broiler production.

General recommendations for the management of nipple systems are:

- 12 birds per nipple. This should be reduced to 9-10 per nipple for birds weighing 2.75 kg or more.
- Nipple height should be monitored daily and adjusted as appropriate. At day old, nipples should be placed at chick eye level. From day 2 onward, while drinking, the back of the chick should form an angle of 45° with the floor.
- Litter, under and around the drinker lines, should be level to allow all birds to have equal access to water.
- Drinker lines should be level to a avoid spillage.
- Individual nipples should be checked regularly to confirm that access is available to birds through 360° (i.e. from all directions). Faulty nipples will reduce birds' access to drinking water. Nipples should be activated and checked by hand before placement to ensure all nipples are working.

- Water pressure should be set according to manufacturers' specifications.
- Nipple lines should be flushed and sanitised weekly.

Bell Drinkers

- When whole house brooding is practiced, a minimum of 6 bell drinkers should be provided per 1000 chicks.
- Drinkers should be distributed evenly throughout the house so that no broiler is more than 2 m from water.
- As a guide to level, water should be 0.6 cm below the top of the drinker until 7-10 days and there should be 0.6 cm of water in the base of the drinker from 10 days onwards.
- The height at which the bell drinkers are suspended should be checked and adjusted daily, so that the lip of the bell is level with the broilers' backs from 7 days onwards.

Agritech

Comparative evaluation of feed conservation in fibreglass and metal silos during summer and winter time

Agritech srl has entrusted the Faculty of Agriculture of the Università Cattolica del Sacro Cuore, in cooperation with Cerzoo, the Research Center for Zootechnics and Environment of Piacenza (Italy) with a comparative study on the performance of some own-manufactured fibreglass silos and other silos in galvanised metal, produced by Chore Time. The study covers a period of time going from July 21st 2008 to February 2nd 2009.

Aim of the Study

The aim of the study was the comparative analysis of the conservation of bulk feed in mealy form stored in fibreglass and galvanised metal silos during summer and winter time.

Compared Samples

1. *Storage silos:* The study was carried out on a total of 6 silos of 6 ton capacity each, supplied by Agritech, 3 of which in fibreglass (VTR) and 3 in metal (MET). The silos were installed in pairs (VTR-MET) in the same environmental conditions with regard to the exposure to the sun.
2. *Stored feed:* The feed selected for the test was commercial compound feed in mealy form, taken from the same production batch, and stored in the silos in the same quantity.

The silos were only loaded by 2/3 of their real capacity in order to reproduce the normal conditions of use in a standard farm, where silos are progressively emptied. To put in evidence eventual effects connected with non-ideal storage and environmental conditions typical of summer, some vegetal oil was added to the main feed as lipidic integration.

3. *Period:* The study was carried out from July 2008 to February 2009.
4. Surveys: During the 4-month summer test following surveys were carried out:
   a) *Temperature (T°):* Using Min. and Max. thermometers, the values of T° MAX., T° MIN. and T° INSTANT in the external environment, the air temperature inside the silos (the empty volume between the cover of the silos and the surface of the feed) and the temperature of the feed (with a thermometer being placed in the first 10 cm of the bulk) were measured on alternate days.
   b) The amount of peroxides released over 20 days, based on feed samples taken from the superior and inferior part of each silo.

   During the 3-month winter test following surveys were carried out:
   c) *Temperature (T°):* Using Min. and Max. thermometers, the values of T° MAX., T° MIN. and T° INSTANT in the external environment, the air temperature inside the silos (the empty volume between the cover of the silos and the surface of the feed) and the temperature of the feed (with a thermometer being placed within the first 10 cm of the bulk) were measured on alternate days.
5. *Results:* As far as the summer period is concerned, it was noticed that:
   a) There are remarkable differences between the temperature of the air inside the silos and that of the stored feed in relation to the building material of the silos. These differences, that arise from the majority of the surveys, are statistically significant, and they are indicated in the charts by some asterisks corresponding to the date of the survey. As one can observe, the large amount of significant marks proves that fibreglass silos can stand thermal fluctuations and control both the air and the feed temperature better than metal silos.

In fibreglass silos (VTR), instant, minimum and maximum temperatures resulted to be better than in metal silos (MET) in the majority of the surveys.

Particularly, the Authors observed higher MAX. T° of the feed in the upper part of the bulk and of the air inside metal silos, as reported in chart No. 3 Delta T° between VTR and MET.

As reported, these differences reach up to over 8°C in the air inside silos, with registered MAX. T° over 45°C in the metal silos. The temperature of feed in the upper part of silos has reached temperatures between 35°C and 45°C in the metal silos, while in the fibreglass silos the highest temperature never exceeded 35°C (this peak was only reached in three surveys). The temperature deltas (Ä) comparatively registered in the temperatures of the feed stored in VTR or MET silos confirm the better performance of fibreglass silos, with a difference in favour of fibreglass up to 7°C.

b) Regarding the amount of released peroxides registered in the upper and lower part of the feed bulk, the fibreglass silos show a lower rate of peroxides and consequently a lower oxidation of the lipids contained in feed than metal silos (...).

Regarding the winter period (November 2008 – February 2009), the Authors had following results:

a) "Max. temperature of the air inside the silos: all temperature data registered in fibreglass silos are statistically inferior to those registered in metal silos. The average difference measured during the 15 test-weeks is about - 106,32%. These values in fibreglass silos tend to match with those of the environmental temperature in the "hottest" weeks, while, similarly to what happens with the minimum temperatures, the lowest ("coldest") values tend to be inferior to those measured in the outside environment".

b) "Max. temperature of the feed inside silos: in the majority of the surveys, the MAX. T° of the feed stored inside fibreglass silos show values that are much inferior to those registered in metal silos.

"Finally, the results obtained in the second part of the test, which integrate and complete the results of the summer test, prove that also in the winter months the

temperature registered in the air and feed inside silos is the parameter which is subject to the most significant variations, and that it largely depends on the building material of the silos. In fibreglass silos, both the temperature of the inside air and the temperature of the stored feed bulk are averagely inferior to the temperatures registered in metal silos".

This e-mail address is being protected from spambots. You need JavaScript enabled to view it

Big Dutchman

ReproMatic and FluxxBreeder – new feeding system especially for broiler breeders. ReproMatic is a feeding system developed by Big Dutchman to ideally meet the particular requirements of broiler breeder management. Only this system allows all birds to receive feed immediately and simultaneously. It combines the advantages of chain and pan feeding. The rugged feed chain is used as conveying system. The feed channel with chain allows for a high filling level and consequently a very high conveying capacity.

FluxxBreeder is Big Dutchman's newly developed feed pan that can be used in rearing for day-old to death production and in broiler breeder production.

The main aim in rearing is for all hens of a flock to reach laying maturity at the same time. A uniform flock can only develop, if all birds have sufficient space to feed. Moreover, pans must be filled simultaneously at the same speed and to the same level to allow all birds of a flock to receive the same amount of feed during restricted feeding.

Features of FluxxBreeder in the rearing phase:

ReproMatic-pullets: The same view with pullets. 16 "true" feeding spaces, which means 60 % more birds per running metre of feeding system as compared to a linear trough;

- 360° flooding mechanism that ensures a high feed level in the pan, especially in the first days of rearing;
- spin-n-lock system allows for simple, one-handed adjustment of the feed level;
- flat pan for an ideal start of day old chicks, a good distribution of feed and reduced feed losses;
- the elevated feed channel and rotatable pan provide the birds with enough freedom of movement;

- eight wings on the outer cylinder of the pan prevent feed losses as lateral feed spillage is not possible;
- the integrated volume reducer allows for small feed rations, thus ensuring fast and simultaneous filling of all pans;
- after the birds have been moved out, the pans can easily be cleaned with a high-pressure cleaner, for drying the pan bottom can simply be opened;
- excess cleaning water or disinfectants can easily drain off through additional holes in the pan bottom;
- the conveying system consists of a feed channel equipped with the Challenger feed chain, thus allowing the transport of large amounts of feed with a high conveying capacity (2 t/h);
- smooth feed saving lip prevents bruises and feed losses;
- sliding shut-off to close individual pans;
- ideal illumination of the pan due to openings in the pan top.

Amacs is Big Dutchman's sophisticated and modern management system for the control of the entire feed supply but also for climate, water supply, bird weighing and lighting. Amacs allows to control a low-maintenance batch. Since the feed lines can be filled when they are suspended there is no need for a large hopper.

Amacs has a modular design and can be used for small and large houses alike, as it can be adapted to the individual situation on a farm. Amacs allows for ongoing, data collection, real-time control and monitoring of individual barns or entire farm complexes – all this from virtually any location in the world.

Chore-Time

Brock Europe offers poultry drinking and feeding systems.

Chore-Time Europe B.V. offers drinker and feeder options for commercial poultry producers.

Chore-Time's popular RELIA-FLOW® Nipple Drinker offers producers a reliable flow rate consistent with the way birds actually drink. This university-tested drinking system features robust, precision-machined, stainless steel parts in the flow-control area of the valve which resist wear and retain their shape for long life and consistent, reliable flow. The system includes a one-piece valve for easy field replacement in retrofit applications. Chore-Time also offers a basic drinker model with its STEADI-FLOW® Nipple Drinker.

For turkey producers, Chore-Time offers the ADVANTI-FLOW® Poult Drinker. The ADVANTI-FLOW Drinker features pockets in the drinker's disc that hold attractive beads of water. The pockets and the disc's scalloped edge work together to control the direction of water flow. The disc also provides leverage for easier triggering of the nipple drinker by young birds.

Dual catch cups with rounded edges for bird comfort maximise water consumption while helping keep floors dry.

Chore-Time recently introduced a new drinking system regulator which is easier for users to install, manage and maintain. Designed for long, reliable service, the regulator requires less pressure to seal than other regulators, resulting in less wear on the sealing mechanism. Simple-to-use, top-mounted knobs activate the "regulate" or "flush" modes for each drinker line, while a bottom knob is used to adjust the operating water column at each regulator.

Also available from Chore-Time is a new, large-diameter, folding stand tube which folds to protect both the tube and the ceiling during house clean out.

The rigid PVC stand tube requires no spring for support, making the water column easier to see and helping the tube to stay cleaner. The large-volume pipe removes more air from the watering system than smaller diameter tubes.

Chore-Time offers poultry producers convenient control of the water pressure levels in all nipple drinker lines in the house with its updated PDS™ Controls. With Chore-Time's PDS (Pneumatic Drinking System) Controls, users can change the pressure in all lines in the house or flush all lines from one remote location.

One of Chore-Time's most popular feeding systems is its Model C2® PLUS Pan Feeding System. The feeder is available in a standard model, as well as models with a shallow pan to help with starting birds or a model with extended fins to enhance feed flow for harder-flowing feed types. The chick-friendly 14-spoke grill design includes feed-saving features to help maximise feed conversion throughout the growing cycle.

Chore-Time's also offers its unique REVOLUTION® Poultry Feeder. Unlike conventional feeder pans with flood windows, the REVOLUTION Feeder includes a Rotary Feed Gate which provides full control of both flood and final feed levels without needing to depend on pan height or gravity to operate. On the floor or in the air, the Rotary Gate gives the REVOLUTION Feeder complete control of both flood and final feed levels.

Chore-Time Europe B.V. is an affiliate of CTB, Inc. Based in Milford, Indiana (U.S.A.), CTB, Inc. is a leading global designer, manufacturer and marketer of systems and solutions for the poultry, pig, egg production, and grain industries. Its products and services are "Helping to Feed a Hungry World®" through improved efficiency and air quality management in the care of poultry and livestock as well as in grain storage, handling, conditioning and drying.

Founded in 1952, CTB has been dedicated to "Leadership Through Innovation®" throughout its history. The company operates from multiple locations in various countries around the world and serves its customers through a worldwide network of independent dealers and distributors.

4

Commercial Poultry Nutrition

The meat and egg industries have undergone remarkable change over the last 30 yr, and it seems that this will continue, albeit at a slightly reduced rate, over the next 10 to 20 yr. The Americas and Asia will continue to meet world demand, and the only real challenge to the dominance of global poultry meat production will come from the swine industry. Broilers and layers in general are remarkably healthy, and we are now seeing unprecedented low levels of mortality. Our current success is due to genetic selection, availability of efficacious vaccines and antibiotics, and a growing awareness of the importance of biosecurity and general farm hygiene. An emerging area of concern in poultry nutrition is the accumulation of Zn and Cu in soil, and this may attract legislation regarding manure composition.

Likewise, there is emphasis on quantitating NH_3 release from many industries including agriculture. We are unlikely to see any major change in nutrient needs of broilers and layers over the next 20 yr. On the other hand, diet formulation, feeding programs, and production goals are continually changing, and these factors affect the work of poultry nutritionists. Our future roles will be governed by the need to accommodate ever-increasing genetic potential, the demand for simple diets devoid of most pharmaceutical products, and the effect of poultry products on human health. Traceability of meat and eggs is inevitable either through legislation or through marketing strategies. Such traceability will require accountability of the composition of poultry feeds. At best, quality control programs at feed mills provide information that is incorporated in a historical database. The feed industry will be compelled to develop real-time feed analysis.

Description of Problem

The poultry meat industry has undoubtedly been the most successful of any of the animal industries, and the egg industry is now making strides in new product development. Today, we see 3-kg male broilers at 40 d of age and 1.7-kg females at 30 d of age, whereas white egg layers are capable of producing in excess of 330 eggs in 52 wk of lay. There is often debate about there being an end point to this increased genetic potential, yet the few geneticists affecting the industry today tell us that selection pressure will be little reduced in the foreseeable future. It is possible that selection criteria may change slightly and that commercial performance will be more balanced with reproduction in meat birds and with skeletal integrity in both broilers and layers.

There is a prediction for consumption of around 65 million tonnes of broiler meat in 2015, which equates to a yearly live production of 80 million tonnes or about 40 billion 2-kg birds. The industry will require some 160 million tonnes of feed. There will likely be a change in location of such production. Currently we have major production centres in the Americas, Europe, and Asia. The Americas and Asia will continue to meet world demand, whereas Europe seems destined to supply its own local market for niche products. Within the Americas, the United States and Brazil will continue to be the major producers and exporters, whereas in Asia, China will likely represent the major region for growth, both for domestic production and also for exports. It is obvious that although it is possible to produce broilers in almost any geographical region, it is the location of ever more sophisticated processing facilities that dictate major centres of meat production. Some countries will continue with marketing of live birds (e.g., Mexico and Peru), and although this effectively excludes imports, it does limit future expansion and profitability and now poses unique problems of biosecurity in movement of live birds. Processing and further processing are the keys to market penetration of poultry meat, and these facilities are becoming increasingly expensive to establish and maintain. The current trend to producing heavier birds is merely a response to reducing processing costs per unit of meat yield. The limiting factor to processing in the next 20 yr will be an adequate and economical supply of water that meets every increasing standard for pollutants and contaminants. We will meet these standards by utilising water purification systems, which will necessarily add to our cost of production.

The only real challenge to the dominance of global poultry meat production will come from the swine industry. As they develop production strategies similar to those characteristic of broiler chicken production, such specialisation and economy of scale will result in cheaper cost of production. However, their higher maintenance energy requirement will always mean a less efficient feed utilisation, and so the broiler industry will continue to lead in least-cost production of lean meat.

Also, as the pork industry delves into further processing, like the poultry industry, they will see advantages to marketing of heavier pigs, and this will help the broiler industry to maintain its competitive edge. It seems likely that the emerging modern swine industry will mirror developments in the poultry industry, and in most situations, our main competitor will be physically situated in very close proximity to poultry producers. The swine industry will also be our major competitor for sourcing of corn, soybean meal, and fats.

The egg layer industries are also undergoing change, in terms of potential for egg processing and more recently the start of major international trade in shell eggs. The major production areas will be the Americas and Asia, with some niche markets relating to housing systems remaining in Europe. Layers, like broilers, will be fed diets based on corn and soybean meal. By 2015, we will have 1.2-kg mature weight white egg strains eating about 95 g daily and producing 345 eggs by 72 wk of age. For brown egg birds, intake will be around 100 g daily, and production will be 335 eggs by 72 wk of age. Although cages will continue to be the main system for housing birds, there will be niche markets for noncage production systems. However, cage systems will be the only competitive production alternative for global export markets, especially for processed egg products.

We have to resolve the ongoing problem of eggshell quality, which considering the number of eggs that will be produced by layers in the future, will invariably become a more serious problem. We need a better understanding of Ca flux of the hen throughout the entire 24-h cycle of egg formation and how this is affected by nutrition, housing, environment, and health status of the bird. There will undoubtedly be new niche markets opening up for eggs and egg products. There is going to be greater emphasis on value-added production of egg products, and for this reason, we need to know a great deal more about factors that influence the solids content of an egg and not merely factors that influence egg weight. We can also have the bird include novel fats and proteins into the egg, which make them even more attractive for human consumption, as novel designer foods for other animals such as baby pigs and other industrial uses.

Effect of Changes in Production Systems

The broiler industry has become more standardised in housing and production systems over the last 10 yr, and in large part, this is a reflection of the need for greater control of the environment for today's meat birds capable of 70 to 75 g of growth per day to 40 d. On the other hand, the egg industry is becoming more divergent in housing systems, a situation often imposed by government regulations regarding welfare, the environment, or both.

Worldwide, there has been general standardisation in the type of housing and management systems of broiler farms, and to some extent, this leads to standardisation of feeding programs. New farms today will likely involve a site for 8 to 10 houses, each holding around 50,000 to 60,000 birds. These houses will be fully controlled environments or open-sided with more elaborate insulated curtains, usually incorporating tunnel ventilation. Although simple pole and wire structures were commonly built in temperate regions in the 1970s, we now realise that the extra cost involved in having greater control over environmental conditions is more profitable in terms of broiler performance over the 20- to 25-yr life span of the building.

In the mid 1980s, broiler producers in tropical climates were encouraged to abandon their low-lying farms because of problems with heat stress and establish new facilities at higher altitudes, which were much cooler. Since then, ascites emerged as a major cause of mortality, and farming broilers of any age at <15°C caused high mortality. Unfortunately, there are always cool/cold nighttime conditions at high altitude, and so these new locations were unsatisfactory.

Therefore, we are once again concentrating on lower altitude sites for broiler farms, because today, the effects of heat distress can be better managed with new environmental systems and a better understanding of the effect of nutrition on heat-stressed birds. Another future concern is an adequate supply of clean water for any broiler operation. In many regions, ground water levels are receding, which means more power is required to pump to surface levels. In arable regions, the levels of minerals and pesticides will become more critical as a consequence of their gradual (20- to 30-yr) movement through soil and our inevitable advances in being able to detect ever-decreasing quantities of any residue. Likewise, the trace mineral content of water is changing over time, and this can affect nutrition.

Bird Health, Antibiotics, and Alternatives

There are currently few major health concerns with either broilers or layers, and we are now seeing unprecedented low levels of mortality. Our current success is due to genetic selection, availability of efficacious vaccines and antibiotics, and a growing awareness of the importance of biosecurity and general farm hygiene. Today, infectious disease is not usually the major cause of mortality or morbidity, even though such infectious agents are ever present on the farm or in the general locale. Our major concerns are more often loosely termed metabolic disorders, the most notable ones listed in order of economic importance being sudden death syndrome, skeletal disorders, and ascites in broilers and cage layer fatigue and fatty liver in layers. Most metabolic disorders can be prevented by growing meat birds more slowly, a situation that is often not economical. If growth control is considered, then systems that affect early development must be considered.

The most striking feature of modern growth potential occurs in the first week, in which today we have a male broiler capable of increasing its BW by 300%. Using specialised prestarter diets, we have actually achieved a 400% increase in the 0- to 7-d growth, yet the 300% achieved with commercial diets is extraordinary compared with other farm animals. Relative growth after this time is little different, yet the additional weight achieved by d 7 is compounded by subsequent weekly percentage increases, such that by 8 wk, we have a broiler that today is 1 kg heavier than its 1970 counterpart. However, tempering this early growth, for reasons not fully understood, does seem to reduce the incidence of metabolic disorders occurring after 28 d of age.

Such growth reduction can be achieved by using mash diets, very low nutrient density diets, perhaps by physical feed restriction, or more commonly by extended periods of darkness. We can temper early growth rate without too much loss in feed utilisation, because during the first 7 d of growth, some 80% of feed is used for growth and only 20% used for maintenance. At 49 d of age, only 20% of feed is used for growth, whereas 80% is used for maintenance, and so trying to temper growth at this age is detrimental to feed efficiency. If mortality/ morbidity can be reduced sufficiently by a system of early feed restriction, then the overall feed efficiency of the flock can actually be improved. There is some indication of compensatory growth in broilers. It will be increasingly difficult to sustain such corrective management procedures to what is effectively a consequence of continued genetic selection for faster growth and increased potential for feed consumption. In the near term, current and emerging metabolic disorders will be of sufficient magnitude to necessitate a moderation in selection pressure for growth rate.

Control over coccidiosis and necrotic enteritis is likely to be our major future concern regarding infectious disease of broilers. With the introduction of ionophore anticoccidials, concerns over coccidial control diminished, because these products were universally efficacious. Resistance to older chemical coccidiostats occurred quickly, because their mode of action involved only a few enzyme systems within the oocyst controlled by a limited number of genes.

The ionophores, on the other hand, caused massive disruption to cell membranes, affecting numerous enzyme systems and control mechanisms. As anticipated, it has taken oocysts much longer to mutate such that oocysts can withstand this more widespread biochemical disruption. Unfortunately, ineffective control of coccidiosis often leads to necrotic enteritis caused by an opportunistic clostridial bacterium. Although coccidial vaccines have been used successfully for breeders, their use to date for broilers has met with variable success, again due to development of resistance. A novel approach in the future may well be the sequential use of a vaccine followed by an ionophore anticoccidial.

Current production systems rely on feedborne antibiotics and growth promoters. With current growth rates of broiler chickens, the classical effects of growth promoters are less easy to quantitate. However, most of these compounds are efficacious against clostridial infection and without them, it is often difficult to control necrotic enteritis and associated outbreaks of coccidiosis. However, it seems as though we are destined to routinely use less of these pharmaceutical products, and although they may be available on prescription for treatment use, alternatives are a fruitful area of research and development. So-called probiotics and prebiotics seem logical alternatives to antibiotics and growth promoters, and the key to their use is very early dosing of the bird. Such treatment of chicks on arrival at the farm may prove to be too late to prevent colonisation of pathogens. This leaves the hatchery as the most logical site for treatment.

Reuse of litter material is another contentious issue. In the United States, 6 to 8 flocks are commonly used on the same built-up litter, whereas in most other regions of the world, litter is removed and houses thoroughly cleaned after each crop of birds. Many producers argue that reusing litter simply does not fit into modern biosecurity and health management programs. However, if the previous crop of broilers was healthy, then their litter and the dirty facilities provide a useful resource for initiating a competitive exclusion program (i.e., the general microbial population from the previous flock is used to seed the next flock of birds). Rather than dismiss the idea of reusing litter, there are circumstances in which it can be used to advantage in a health management program.

Environmental Issues

Until recently, the poultry industry has been unaffected by environmental issues that have arisen with some other intensive animal industries. Broiler manure (litter) is relatively easy to handle and does not have the same odor problems usually associated with intensive swine farms and beef feedlot operations. However, the sophistication and scale of poultry processing facilities today means that broiler farms are necessarily located within reasonable trucking distance of this central unit. This intensification of bird numbers inevitably leads to greater challenges for proper disposal of manure. The current and future problems are somewhat predictable based on the past experiences seen by the animal industries in Holland and Taiwan.

We can certainly reduce N and P output in manure, simply by limiting the dietary inputs. In many situations today, it is still most economical to overformulate for these 2 nutrients, and imposing maximum constraints during formulation often leads to increased cost. The impetus to reduce N and P loading of manure will necessarily come from legislation involving nutrient management of farmland. In these situations, P is likely to be the major problem, because in many areas today, we are at the saturation point in soil levels of P as dictated by current environmental guidelines. Corn and soybeans take minimal quantities of P from the soil, suggesting that conventional cropping is not a short-term solution to any future problems. The P levels in soil can most quickly be depleted by growing forages, and removing this as hay or silage, a situation that is not always convenient in areas of intensive poultry production. We can certainly reduce the P output of broilers through use of phytase enzyme added to feed, although it is unfortunate that conventional pelleting temperatures and formaldehyde treatment of feed deactivates some phytase enzymes.

Low-phytate content corn and soybean meal will certainly help to limit P excretion.

Incineration may be an answer in many locations. With efficient incineration, only inorganic elements remain, and these can be used as a fertilizer or as an animal feed ingredient. The ash remaining after incineration is much condensed in volume (95%) and can be more economically moved away from the immediate area of production, to locations where perhaps the corn and soybean used to feed the birds was grown. Intensive broiler production does not create P or N; it merely concentrates its presence, and redistribution back to its original source seems the ultimate example of recycling.

An emerging area of concern in poultry nutrition is the accumulation of Zn and Cu in soil, and again, this may attract legislation regarding manure composition. Likewise, there is emphasis on quantitating NH_3 release from many industries including agriculture.

Nutrition And Feed Programs

There will be no major change in nutrient requirement of broilers and layers over the next 20 yr. In fact, nutrient requirements of these birds have changed very little over the past 50 yr. On the other hand, diet formulation, feeding programs, and production goals are continually changing, and it is these 3 factors that affect diet specifications. Our future roles will be governed by the need to accommodate ever-increasing genetic potential, the need to grow birds on simple diets devoid of most pharmaceutical products, and the effect of poultry products on human health.

If we want meat birds to achieve ever-increasing weight for age, or more specifically, decreasing age for weight, then greater emphasis will have to be placed on both early- and late-phase nutrition. With females that are potentially marketed at 30 d of age (1.75 kg of live weight), the conventional starter period of 0 to 15 d represents 50% of growout time, and birds must be on a withdrawal diet by 24 to 25 d. The concept of specialised prestarter diets becomes more interesting, whereas the philosophy around producing inexpensive withdrawal diets will become a dated concept.

In using prestarters, one assumes that there will be faster early growth (or more uniform growth) and that this will translate into greater weight for age. Each gram of additional weight at 7 d of age translates to 5 g of extra weight at 49 d. Therefore, a chick at 180 vs. 150 g at 7 d can be expected to be 150 g heavier at 49 d. Formulating prestarter diets revolves around the selection of digestible ingredients rather than there being a need for higher nutrient density. Prestarters either precondition the chick such that it can digest complex substrates or provide more digestible substrates until the enzyme production of the chick has matured. Although corn-soybean meal diets are regarded as ideal for poultry, there is evidence that digestibility is suboptimal for the young chick. The idea in formulating prestarter diets is to correct any such deficiency and so hopefully increase early growth rate, improve uniformity of such early growth, or both.

Two types of prestarter diets are used for broiler chickens. The first option is to use greater than normal levels of nutrients, whereas the alternate approach is to use more highly digestible ingredients. If we increase nutrient supply by 10 to 15%, it should be possible to correct any deficiency in digestibility and so realise expected AME_n and amino acid utilisation. A potential problem with this approach is the acceptance that nutrients will not be optimally digested. Unfortunately, such undigested nutrients will fuel microbial overgrowth. An alternate approach is to use more highly digestible ingredients, with little change in level of nutrients. Such prestarter diets are very expensive, because alternative ingredients are invariably more expensive than are corn and soybean meal per unit of available nutrients. Using these ingredients, it is possible to achieve 200 g of BW at 7 d, compared with 160 to 170 g with conventional corn-soybean diets. This improved early growth rate usually ensures a heavier bird at subsequent ages.

Toward the end of the feed program, we will have to pay greater attention to nutrient needs during the last 5 to 10 d of growout. There is currently considerable interest in defining amino acid needs of older broilers. At some time, we will likely have to consider programs that temper the growth rate of broilers. This concept will initially be applied to the small weight category female, marketed at around 1.75 kg of live weight. Currently, 32 to 34 d is a common market age, although with continued genetic progress, market age could be 28 d in 5 to 10 yr time. At 28 d, the limit to successful production will be skeletal integrity of the broiler. The skeleton of such young birds will not likely be adequately calcified/mature to withstand the rigours of commercial processing. Either we have to rethink nutrition/feed management or we have to temper growth. It is very difficult to reduce growth rate simply by altering nutrient density. With high-energy or low-amino acid diets, the birds become obese, whereas with low-energy diets, the broiler still has an amazing ability to increase its feed intake to maintain energy intake. For broilers, the limit to growth will be the maturity of the skeleton, relative to bird welfare and the stresses of mechanical processing. For layers, the limit to production will be eggshell quality. Metabolic disorders related to skeletal integrity will be our major challenge over the next 10 to 15 yr and will ultimately pose the limit to increased productivity in all areas of poultry production.

Human Health, Traceability, Accountability

The effect of animal production and animal products on human health is now of paramount concern to governments. In many countries, health care accounts for around 40% of government budgets. Consequently, there is usually support for modifying poultry products that positively affect human health. For the animal and food industries, functional foods represent an opportunity for increased revenue. Concurrently, modified animal products provide consumers with food choices that are healthier or more appealing. Function foods have been in existence for many years, although it is only in the last 15 yr or so that they have become more mainstream and available through the larger retail outlets. Omega-3-enriched eggs were the first such mainstream functional food, and in addition to capturing a sizeable portion of the egg market, they have spurred the development of similar animal products.

Value-added functional foods can be produced through the modification of the conventional production systems and usually by modifying the diet of the bird. For the most part, such approaches are straightforward, and the outcome in terms of product composition is fairly predictable. However, as society moves to accepting/demanding more processed food, the role of poultry nutrition in contributing to functional foods will become less clear. During the processing of food products, it is perhaps easier to add or enrich the nutraceuticals directly into a product rather than rely on the bird to bioconvert and deposit within the body or eggs. Currently, functional animal foods revolve around the manipulation of fats and fat-soluble nutrients. It is fairly easy to manipulate the total quantity of fat, and its fatty acid profile, in both eggs and meat. Protein, amino acids, and most minerals on the other hand are very resistant to manipulation through dietary modification.

An often-neglected component of functional foods is the emotional value of food. In many instances, the purchase and selection of foods is based on emotional issues, and in reality, these decisions relate to the emotional well being and health of the consumer. In this regard, animal welfare is often a key issue, and certainly, organic certification usually involves welfare standards. In the future, we may well see environmentally friendly foods being marketed, especially in view of traceability, and this will again affect poultry nutrition, especially in terms of oversupply of certain nutrients.

Traceability of animal products is inevitable either through legislation or through marketing strategies. Such traceability will require accountability of the composition of poultry feeds. Our current quality control procedures are a major weakness in the food supply chain. Rapid analysis systems are being developed in many industries, but at best, most analytical systems at feed mills provide information that is incorporated into a historical database. The feed industry will be compelled to develop real-time feed analysis. Such analyses will be for components that affect poultry performance/economics, such as protein and energy, whereas other analyses will be required for bacteria, pharmaceutical products, and contaminants.

Various light and sound reflectance/absorbance techniques will be developed to achieve these goals. We currently have infrared analysis, and in the future, use of directed lasers and radar may be used in analysis. Although near-infrared analysis has been used for 30 yr in feed analysis, there are still many unknowns to using the technology, and unfortunately, little research is carried out in this area. Within 20 yr, it is conceivable that we will have infrared, laser, or sound-based detectors placed in the surge bins feeding the mixer. Using fibre optic probes attached to a central analyser, it should be possible to provide immediate analysis of ingredients as they flow into the mixer, and this information will be integrated with the formulation program and the weigh scale. Directed fine-tuning of the formulation and weight of each ingredient should ensure more consistent composition of a diet. Such nutrient analysis would also be stored and used to update standard values and so provide a more precise estimate of current ingredient usage. Real-time analyses for pharmaceuticals/ contaminants obviously imposes an even greater order of complexity in analyses, yet nutritionists and feed mill managers are encouraged to keep themselves appraised of the rapidly advancing disciplines of chemical, physical, and biological analyses.

Ammonia and Particulate Matter Emission Factors for Poultry Operations

Limited data on pollutant emissions from poultry operations are available to assess the effect of these operations on the environment and to put their contribution in perspective with other sources of pollutants. To alleviate this problem, numerous studies at various poultry facilities have been undertaken to improve the knowledge base in quantifying emissions of NH_3 and size-fractionated particulate matter (PM). For these emission data to be of practical use for government agencies and policy makers, the emission rates must be reported as an emission factor with a production unit that enables the emissions from one poultry operation to be correlated to another poultry operation. This paper presents a compilation of NH_3 and PM emission data from several studies in the form of emission factor on a per-500 kg of live weight or animal unit basis. In addition, best management practices that lower pollutant emissions from poultry operations have been reported along with their effectiveness at reducing NH_3 and PM. Unfortunately, the compiled data were insufficient to characterise the variability in emissions caused by differences in house design, suggesting that more studies are needed to complete a comprehensive emission inventory. Once complete, this inventory will enable poultry producers to estimate emissions from their facilities and, if necessary, select management practice(s) to lessen their emissions of NH_3, PM, or both.

Description of Problem

Government agencies and research groups have initiated programs aimed at improving the understanding of pollutants released from smaller area sources such as livestock housing operations. Livestock housing is believed to contribute significantly to regional air pollution by emissions of NH_3 and particulate matter. In many regions, particularly those in North America, concentrated animal feeding operations are exempt from environmental regulations that industrial sources fall under. As a result, the emissions from animal housing operations are poorly characterised and often estimated by extrapolating emission factor (EF) data from limited data sets.

One type of animal housing that has relied heavily on extrapolated EF is the commercial poultry sector. In many North America jurisdictions, a single EF for a given pollutant has been used to describe the emissions from all types of poultry production. For this reason, several research groups have placed a higher priority on improving EF for poultry over other types of livestock housing.

The production of poultry results in the release of several pollutants to the atmosphere, although most research has focused on NH_3 and particulate matter (PM) emissions because of their detrimental effects to human health and the environment. Each type of poultry operation generates different amounts of these contaminants, and their emission rates are further confounded by the different house management and operation practices. Information on the emissions of NH_3 and PM from different types of poultry production is currently available, but the various findings need to be summarised by translating them into a comparable EF format using a consistent production unit.

This compilation of comparable EF will allow any knowledge gaps to be identified that may warrant additional measurement studies. A comprehensive set of emissions data will also enable individual poultry producers to select the most appropriate EF to estimate the target pollutant emissions from their specific facility. Government agencies can benefit from these data by developing more comprehensive inventories of NH_3 and PM releases. However, for policy makers to truly benefit from these data, they must be educated with additional information on mitigation techniques aimed at lessening the effect on air quality.

Various research groups have been investigating several innovative best management practices (BMP) for poultry facilities to reduce their effects on both the indoor and outdoor environments. For these BMP to be useful for poultry producers and government agencies, their potential to reduce the emission of a target pollutant must be reported for each type of poultry operation and house design. This information would enable policy makers to begin coordinating with poultry producers to work toward feasible pollutant reduction programs. Although this summary only investigates the potential pollutant reductions of the BMP, additional considerations, such as capital costs, availability, and maintenance costs of the BMP, need to be assessed before wide-scale implementation.

The objectives of the paper are to report the NH_3 and PM EF for various types of poultry operations and house configurations, which can be used to determine whether or not the variability in emissions has been fully characterised. The paper also describes several BMP along with their potential effectiveness at reducing NH_3, PM, or both, emissions that will provide an indication of the current knowledge gaps in this area.

Emission Factors

An EF production unit or activity level is usually selected at the discretion of the individual(s) reporting its value. In regards to poultry production, EF are commonly expressed with units of per house, per

1,000 birds, per animal unit (AU; equivalent to 500 kg of live weight), or per heat-producing unit. Because this summary addresses all sectors of the poultry industry, the AU basis is most appropriate for comparison between different housing systems and bird weights.

For studies that do not provide the necessary information to convert their reported emissions to an EF on an AU basis, the following information provided by the NFPC for typical poultry production was assumed as default for the industry:

For broiler chickens, an average of 40,000 birds per house is raised in a single production cycle that lasts 45 d with 2 wk allotted after each flock to clean and disinfect the house. This equates to 6 production cycles annually. Considering the length of a production cycle, the average weight of a broiler (including both male and female statistics) is 1.56 kg per bird.

Layer chicken facilities operate continuously throughout the year with an average of 20,000 birds in the poultry house. The average weight of a layer is 1.58 kg per bird.

Turkeys are raised in a similar manner to broiler chickens with an average of 6,000 turkeys per house in a single production cycle lasting 20 wk. On average, turkey facilities are empty for 4 wk per year for biosecurity reasons. Given this time frame, only 2 complete flocks are raised annually with a third underway. During a growth cycle, the average weight of a turkey is 6.3 kg assuming an equal ratio of male and female birds.

It should be noted that these statistics are most appropriate for the Canadian poultry industry and may not reflect common practices in other jurisdictions such as the United States. Size-fractionated particulate matter emission factors (EF) for different types of poultry operations, ventilation systems, and manure management systems

Emissions

Ammonia emission factors, based on equivalent 500 kg of live weight AU, are summarised in Table 1Ñ!. Ammonia emissions from poultry houses vary significantly based on the type of poultry operation, their climatological region, and the management practices used by the house operators. For this reason, the geographical location, type of ventilation, and manure management system were identified with the NH_3 EF summarised.

The NH_3 EF for broiler production ranges, on an average basis, from 57 g/d per AU up to 391 g/d per AU, although most are consistently between 150 and 225 g/d per AU. This range of EF is only slightly higher than the 100 g/bird per year (equivalent to 118 g/d per AU using the NFPC statistics) proposed by the US Environmental Protection Agency.

The highest EF for this type of operation was developed by Wheeler et al. and Lacey et al. based on data from broiler houses in the United States (Kentucky and Pennsylvania and Texas, respectively). It is a common practice at commercial broiler facilities in most parts of the United States to reuse the floor litter for up to 3 or 4 bird-production cycles before replacing it with fresh bedding. In other regions, such as Canada and Europe, the litter is typically replaced after each production cycle.

This manure management practice may explain the elevated NH_3 emissions in the United States. Because some inconsistencies in emissions still exists among the EF developed from the European studies using similar manure management practices, the ventilation system of the poultry house may be a better indicator of the EF variability. Only a few studies provide information on the type of ventilation system used at the studied facility, which is insufficient to confirm its influence on the emission of NH_3.

There is a large degree of spread in the developed EF for layer chickens, which range, on an average basis, from 64.8 g/d per AU up to 468 g/d per AU (or 724 g/d per AU if the variability in the Heber et al. study is included). It is difficult to identify definitive reasons for some of the variability, because there is a variety of manure management systems used in egg production, and very little information is given on the type of ventilation system. Because NH_3 emissions typically result from the decomposition of poultry manure, layer operations that use a battery cage system with manure removal belts frequently generate less NH_3 emissions than those without a manure removal system (i.e., percheries and deep-pit systems). Once again, the United States studies consistently reported higher NH_3 EF for layer hens than those conducted in Europe.

Three studies have quantified the NH_3 emissions from turkey production, but 2 studies did not report the type of turkey operation. The NH_3 emissions from the unspecified turkey production (126 and 113 g/d per AU) are comparable to the grow-out turkey facility studied by Gay et al. (120.5 g/d per AU). In addition, these EF are similar to those for broiler chickens on an AU basis. The brooder facility studied by Gay et al. emitted much less NH_3 (7.2 g/d per AU). This

EF may be misleading, because the emissions were measured for 2 d only with the turkeys being introduced to the facility shortly before the start of the measurement campaign. The turkeys may not have excreted enough fecal matter to generate representative EF.

PM Emissions

Several characteristics of PM can be used to describe its toxicity, including its size (or aerodynamic diametre, AD), origin, formation mechanisms, chemical composition, and pathogenicity. Because the AD of a particle will determine if it is capable of penetrating deep into the human respiratory-cardiopulmonary system, it is the most commonly used particle classification scheme.

In the past, particles have been divided into 3 size classifications: total suspended particulates, those capable of entering the respiratory tract (or particles with an AD equal to or smaller than 10 μm, PM_{10}), and respirable dust (RD; equivalent to 5 μm or less). More recently, 2 additional particle sizes have been introduced to better describe the toxic effects of PM, specifically particles with AD equal to or smaller than 2.5 and 1 μm ($PM_{2.5}$ and PM_1, respectively). Limited research has been conducted at poultry facilities to quantify the emissions of the latter 2 particle size classifications, and as such, the developed EF may not accurately represent the average PM emissions from typical poultry houses.

The amount of total suspended particulates emitted on an AU basis by broiler chickens (170.2 g/d per AU on average) is significantly greater than that released by layer chickens (42.8 g/d per AU on average). However, the PM_{10} emissions from both types of operations, as well as turkey production, are all within the range of 0.23 to 15.8 g/d per AU, which implies that a large portion of the total suspended particles are in the larger size fraction. Possible sources for the coarser particles are mechanical generation by the poultry, the manure management systems, and the feed delivery systems. Contrary to the fairly consistent PM_{10} EF reported for all types of poultry facilities, the RD EF are larger for broiler operations than all types of layer operations. In fact, the RD emissions, a smaller size fraction than PM_{10}, from broiler houses are greater than the measured PM_{10} emissions from similar houses. Given that the studies reporting RD emissions were conducted 5 yr earlier than those reporting PM_{10} emissions, it is possible that the instrumentation used to quantify the PM emission rates have improved the EF estimates. Advancements in PM measurement technologies have enhanced the on-site and semicontinuous quantification capabilities of the instruments, which has eliminated the inherent errors associated with sample transport and laboratory analysis.

Since recognising the importance of the 2 fine PM size classifications, only 2 studies have generated poultry house EF for $PM_{2.5}$. The 2 studies, one at a layer facility and the other at a broiler house, reported similar $PM_{2.5}$ EF on a live weight basis. The mean $PM_{2.5}$ release for both types of operations is slightly greater than 1 g/d per AU. The broiler house study is the only one to report a PM_1 EF, which was estimated to be slightly less than 1 g/d per AU. The nearly equivalent $PM_{2.5}$ and PM_1 EF reported indicates that the fine fraction of PM emitted from poultry operations is mostly comprised of submicron particles. However, this conclusion is not definitive, because it is difficult to gauge variability within the poultry sector with only 1 PM_1 EF.

5

Feed Manufacturing Technology: Current Issues and Challenges

Grain is ground prior to mixing to increase surface area for improved rate of digestion, decreased segregation and mixing problems, and to facilitate further processes such as extrusion or pelleting. Extensive grinding requires more energy; however, even small improvements in feed efficiency will often justify the added cost (Ensminger, 1985).

A study to evaluate the effects of grain sorghum particle size on nutrient digestibility was conducted by Owsley et al. (1981). Using ileal cannulated pigs, they found that upper tract digestibilities of N, DM, GE, starch, and most amino acids were increased as particle size was reduced. Total tract digestibility trends were similar but the differences were not as great. For example, ileal starch digestion was 19% greater in pigs fed grain sorghum ground through a 3.2 mm screen vs. those fed a coarsely rolled grain sorghum. Total tract difference was only 3%. It is recognised that starch is fermented to VFA in the large intestine and absorbed by the pig (Argenzio and Southworth, 1975). However, synthesis of VFA into glucose is less efficient than release and absorption of glucose fromstarch in the small intestine (Black, 1971). It can then be argued that decreased particle size of grain is one technique useful in improving its utilisation in growing finishing pigs. Reducing the particle size of barley by 14% (789μ vs. 676μ) improved ADG and G/F by 5% for starter pigs (Goodband and Hines, 1988). The finely ground barley gave G/F equal to that of grain sorghum ground to 753μ. They concluded that the nutritional value of barley could be improved more than grain sorghum by fine grinding. Ohh et al. (1983) found improved G/F and DM, GE, and N digestibility when the particle size of corn and grain sorghum was reduced. Roller milling and hammermill grinding gave similar results. Wu (1985) found a response to fine grinding of corn in finishing pigs but not in weanling pigs.

Performance of pigs fed wheat may not be improved by extensive particle size reduction. Average daily gain and G/F were similar in starter pigs fed diets containing wheat ground to 860μ or 1710μ average particle size (Seerley et al., 1988). Growing pigs fed wheat ground to the same particle size were 8% more efficient than pigs fed diets with 1710μ wheat but ADG was not affected. Finishing pigs gained 9% slower and were 6% less efficient when fed diets with wheat ground to 860μ vs. 1710μ. These studies would indicate that optimum particle size of wheat increases for older pigs.

Fine grinding has been implicated in the development of ulcers in the esophageal region of pigs (Reimann et al., 1968; Hedde et al., 1985). This is still an area of active investigation. Cabrera et al. (1993) found that stomach morphology was negatively affected by fine grinding (<600μ) of corn and two grain sorghum genotypes; however, improved performance would likely make fine grinding an acceptable compromise. They also found a dramatic decrease in daily DM and N excretion as grain particle size was reduced.

Wondra et al. (1993) investigated the influence of mill type (hammermill vs. roller mill) on finishing pig performance and stomach morphology. Mill type did not affect growth performance, but pigs fed corn ground by a roller mill had greater digestibilities of DM, N and GE and excreted 18% less DM and 13% less N than pigs fed hammermill ground corn. Pigs fed the roller mill ground corn had slightly better ulcer scores than those fed hammermilled grain.

Information regarding the effect of grain particle size on broiler performance is limited. Reese et al. (1986) reported improved G/F for broilers fed diets with corn ground to 910μ vs. 1024μ but only when the diets were pelleted. The authors implied that uniformity of particle size might also affect broiler performance. Healy et al. (1994) investigated the effect of particle size of corn and hard and soft grain sorghum on growth performance and nutrient utilisation in broiler chicks. Starting with seven-day-old chicks, they conducted a 21-day growth assay using diets made from each grain ground to 900, 700, 500, or 300μ. Gains for day 0 to 7 were maximised at 500μ for corn, 700μ for the hard endosperm grain sorghum, and 300μ for the soft grain sorghum. Maximum G/F for corn and soft grain sorghum was 300μ and 500μ for hard grain sorghum. For days 0 to 21 gain was optimum for corn at 700μ, for hard grain sorghum at 500μ, and for soft grain sorghum at 300μ. The authors concluded that reduced particle size improved growth performance and that, if properly processed, grain sorghum has a nutritional value equal to corn.

The fineness of grind found to be optimum in the above investigation is much finer than typically found in the U.S. broiler industry. In general, hammermills are equipped with 10/64" (4 mm) or 12/64" (4.8 mm) screens, usually resulting in an average particle size in the 800μ to 1000μ range. An interesting finding in the Healy et al. (1994) study is that gizzard weights were less in chicks fed the 300μ treatment. It is apparent that if the organ is not utilised for its intended function, development can be inhibited. This may have a significant economic impact by reducing the proportion of lower value parts.

Current Issues – Grinding

There is a great deal of interest in evaluating roller mills as an alternative to hammermills. The largest feed mill in the world has but one small hammermill used to grind "overs" of the ground grain scalper. All other grinding is done with roller mills.

Some U.S. manufacturers are looking at the common European practice of post-mix grinding to improve pellet quality. In most cases, minor improvements are noted but at a substantial cost in energy and capital. A relatively new concept in mill design has been proposed by a large engineering firm (Ibberson, Minneapolis, MN). In effect, this is a hybrid between pre-grind and post-grind and allows management to select the operation of choice depending upon the operation and feed quality desired.Air-assisted hammermill systems continue to gain acceptance. In addition to increasing throughput by up to 25%, moisture shrink and product temperature rise are substantially reduced.

A new hammermill design has been developed by Buhler, Inc. (Minneapolis, MN). The design is novel in that the shaft is vertical rather than horizontal. Screen changing is easily accomplished and particle uniformity and grinding efficiency is said to be improved over conventional hammermills. Evaluating the effect of grind fineness on animal performance continues to be an active area of research. Much needs to be learned regarding other cereals as well as protein meal in this regard. It is known that fermentation rate is increased as particle size is reduced (for ruminants). Little is known of the effects of fine grinding on hindgut fermentation, as in horses.

Feed (Nutrient) Uniformity And Its Effect On Animal Performance

Mixing is considered to be one of the most critical and essential operations in feed manufacturing regardless of whether it's on-farm or in a commercial facility. Lack of proper mixing can lead to reduced diet uniformity, affecting not only animal performance but regulatory compliance as well.

Few feed manufacturers really know how their mixers are performing or what mix time is required for a given diet. An ongoing survey of commercial mixers found that over 50% did not meet the "de facto" industry standard of a coefficient of variation (CV) of less than 10% when using methionine and lysine as the tracer (Wicker and Poole, 1991). A survey of farm feed mixers found very similar results (Stark et al., 1991). In their study, 42% of the participants had CVs of less than 10%, 47% were between 10 and 20% CV, and 11% had CVs greater than 20%. Salt was used as the tracer in this study.

Even though the importance of diet uniformity is intuitive, there is very little credible research that relates diet uniformity to animal performance. Numerous authors have cited uniformity as one of the most important aspects in feed production (Beumer, 1991; Ensminger et al., 1991; Wilcox and Balding, 1986).

However, credible animal studies relating factors such as mix time, diet uniformity, and ingredient segregation are not available in abundance. Duncan (1973) provided insight into the effect of ingredient nutrient variability on animal performance and quality control. If nutrient density is highly variable, it would have the same effect as nutrient variation because of poor mixing or segregation. McCoy et al. (1994) conducted two experiments to investigate the effects of diet nonuniformity, caused by inadequate mixing, on the performance of broiler chicks. In both experiments, a common diet formulation was mixed for different times to represent poor, intermediate, and adequate uniformity. In the first study, only G/F was improved as diet uniformity increased. In the second experiment (a 28-day growth assay), improvements were noted in ADG, ADFI, and G/F as diet uniformity increased from poor to intermediate. No further enhanced performance was noted as diet uniformity was increased to adequate.

In a recent study, Traylor et al. (1994) examined the effect of mix time on diet uniformity and subsequent growth performance of nursery and finishing pigs. For Experiment 1, 120 weanling pigs were used in a 27-day growth assay. Treatments were mix times of 0, 0.5, 2.0, and 4.0 minutes for a Phase II nursery diet. They found that ADG improved by 49% and F/G was decreased by 16% as diet uniformity was increased. In the finishing pig study, 128 animals were fed from 124 lbs to 260 lbs final body weight. Treatments were the same (mix times) as used in the nursery study. Increasing mix time reduced diet variability (CV) from 54% to less than 10% but did not significantly affect ADG, ADFI, or F/G. However, pigs fed the most nonuniform diet had the poorest rates and efficiencies of gain. Conclusions reached in both of these studies (McCoy et al., 1994; Traylor et al., 1994) were that body size and, therefore, daily food intake, of young animals is a critical aspect relating diet uniformity to animal performance. Older, more mature animals that consume larger meals and retain digesta longer are less sensitive to variation in diet uniformity than young animals.

Current Issues - Diet Uniformity

There is obviously a great deal of research needed to document more fully the level of uniformity needed to optimise performance of target animals. It is disturbing to contemplate the fact that the basis for all formulation, nutrition, and regulatory control is the assumption of nutrient or additive uniformity, yet very little real data exist to support this assumption.

There is a distinct possibility that regulatory agencies such as FDA/CVM will require that all mixers used to manufacture medicated feed will have to be validated as to blending ability and required mix time. At present, at least one Canadian province requires biannual testing of medicated feed mixers. The recent introduction of short cycle mixers will have a great impact on mill design and operation. If the mixing cycle is reduced from 5-8 minutes to 1-3 minutes, it is obvious that other operations, such as hand weighment, batching, and liquid additions, will have to be streamlined. Management will become a greater challenge. Perhaps the most needed item is the development of a dependable mixer testing procedure that is acceptable to both the industry and regulatory agencies. At present, there is no "standard" procedure available that meets the requirements of accuracy, safety, expense, and utility needed for acceptance. This is fertile ground for creative research.

Pellet Quality Issues

Since its introduction in the 1930s, pelleting has become an important process to the feed industry. Estimates vary on the percentage of annual feed production that is pelleted but it is likely in the 60% range. In recent years there has been a dramatic increase in pelleted feed tonnage because of the rapid growth in hog integration. This industry, like the broiler industry, has concluded that the cost of pelleting is more than offset by improved animal performance. Pelleted diets can affect animal performance in a variety of ways.

The above factors are critical for feeding food producing animals. In the case of pelleted feed for horses, the criteria to be applied are somewhat different. Perhaps most important, however, is the ability to prevent selective feeding. Nearly as important, however, is the value of clean, bright feed to the horse owner. These owners typically place a high value on packaging and "presentation," and pellets do have a role to play in this regard.

In addition, pelleting allows the use of a wider variety of ingredients without obvious changes in the physical properties of the diet. This often will allow diet costs to be reduced with little or no affect on performance. Research conducted in Europe and North America has shown that pelleted pig nursery diets will increase ADG and G/F by 9 to 10%. Pelleted growerfinisher diets result in an increase of 3 to 5% in ADG and 7 to 10% in G/F.

It is accepted that pelleting will improve performance; however, the effect of pellet quality and/or diet fines content on animal performance is poorly understood. Stark (1994) conducted two nursery experiments and a finishing experiment to determine the effects of meal vs. pellets and the effects of fines on pelleted diet performance in pigs. In the nursery studies, Phase II diets were fed for 28 days. In the first study, diets were fed as meal, screened pellets, or screened pellets with 25% fines to reflect a low quality pellet. Pelleting improved gain to feed by 12% compared to the meal diet. There was a trend toward reduced G/F ($P < .07$) if fines were included in the pellet diet.

In the second nursery study, similar treatments were used except fines were included at 15 or 30% rather than 25%. Pelleting increased ADG and G/F by 8% and 15%, respectively, compared to the meal diet. The addition of fines did not affect ADG or F/G ($P < .04$). In the finishing study, 80 gilts were used to evaluate the effect of diet form and pellet fines on growth performance. The treatments were 1) meal; 2) screened pellets; 3) pellets with 20% fines; 4) pellets with 40% fines; or 5) pellets with 60% fines. Animal performance was not affected by diet form (meal vs. pellet). The presence of fines in the pellets tended to decrease feed efficiency ($P < 0.09$) but did not influence ADG.

In another recent finishing pig study (Wondra et al., 1994), pelleting improved ADG by 4% and G/F by 6%. Feed intake was not affected. Nearly all broiler and turkey feeds are pelleted. It is well established that pelleting improved both ADG and F/G compared to meal diets. However, recent interest in the broiler industry centres on the effect of pellet quality and dietary fines on bird performance. Scheideler (1991) found that broilers fed 75% pellets (25% fines) had substantially better feed conversion than birds fed 25% pellets (75% fines). Selective feeding and nutrient density changes because of selective feeding were noted during the study.

Turkeys appear to be more sensitive to pellet quality and fines. Several studies have indicated that pellet fines decrease turkey performance (Proudfoot and Hulan, 1982; Salmon, 1985; Moran, 1989; Waibel et al., 1992). Moran (1989) showed a decrease in growth and performance when reground pellets were fed, compared to whole pellets. Proudfoot and Hulan (1982) found decreased turkey performance as fines were increased from 0 to 60%.

Current Issues - Pelleting And Pellet Quality

Perhaps the most challenging issue is the development of the high shear conditioning concept (expanders). Many European manufacturers have incorporated expanders in their pelleting systems over the past decade. Over the past two or three years, the U.S. industry has begun to investigate high shear conditioning as a way to improve pellet quality, mill throughput, feed sanitation, and, ultimately, animal performance. Peisker (1993a,b; 1994) has presented data on the applicability of high shear conditioning. In a broiler feeding study (1993b), the author reported that ME was increased by expander conditioning by nearly 5% in a broiler diet. Significant improvements were found in fat and starch digestion. In an earlier report (Peisker, 1993a), the effect of expansion on starch gelatinization was demonstrated. In general, expansion conditioning doubled the level of gelatinization in feed. The use of high shear conditioning may have significant application in some animal applications; however, because of capital and operating and maintenance costs, the industry must use judgement in identifying those applications. At present, there are as many as one hundred of these systems on order, under installation, or installed in the U.S. Most are in the poultry industry. There is concern that many of the decisions being made are coming from a "crowd mentality" rather than being justified by factual information under domestic conditions. This is certainly an area that needs further unbiased research. Another contemporary research topic is the elucidation and control of factors that dictate pellet quality. Formulation has a tremendous impact on pellet quality; however, it is not yet possible to include "pelletability" factors in least cost formulation programs. Pellet quality issues can be partitioned into several individual components and their contribution of each component. These are: formulation (40%); grind (20%); conditioning (20%); die selection (15%); and cooling/drying (5%). It should be noted that 60% of pellet quality is determined before the feed reaches the pellet mill.

There is significant ongoing disagreement within both swine and poultry integrator companies concerning the affect of pellet fines on animal performance. Traditionally, most broiler integrators have held that there is no advantage of having more than 60% pellets (40% fines) in the feeder pans. Recent privately funded studies, however, have shown that contemporary broilers need at least 80% pellets for optimum performance. In general, swine integrators are finding similar results.

There are actually two issues that must be addressed if a high percentage of pellets is to be presented to the animal. First of all is pellet quality. If a high quality, durable pellet is not produced, no amount of gentle handling will assure a high percentage of pellets at the feeder. If the earlier quality allocation is accurate, we must begin to formulate diets with pellet quality in mind. To date, reliable attributes are not available to use least cost formulation to predict pellet quality. However, there are several charts available that address the relative "pelletability" of most common ingredients. The second issue is screening operations after cooling. The trend in the U.S. for the last 10 to 15 years has been to build feed mills with no pellet screening equipment. Even in older feed mills, the screens are often by-passed or blanked with sheet metal. The net result is that all fines generated within the feed mill are not re-pelleted but are sent to loadout or bagging and carried to the farm. Given the size and capacity of today's feed mills, fines generation is unavoidable. Fines contribute to pellet cooling problems, fat or molasses coating problems, segregation, and selective feeding.

One of the easiest and most effective ways to reduce fines delivered to the feeder is to prevent them from leaving the feed mill by screening. Screening operations are not very expensive in terms of capital equipment or operating cost, but if recycle is significant, a reduction in plant throughput equal to the recycle tonnage can be expected. This brings the discussion back to initial pellet quality. If high quality pellets are produced, recycle will be minimised. Beyond the expander, there have been several recent innovations in pelleting. The first is known as the compactor. This device replaces the standard conditioner and serves a purpose similar to the expander. The compactor will not result in the high levels of gelatinization found with extrusion or expansion but does result in substantial improvements in pellet quality and feed hygiene.

Another novel approach is the Universal Pellet Cooker (UPC) introduced by the Wenger Corporation (Sabetha, KS). This device incorporates long-term conditioning with short-term/high temperature extrusion. The product from the machine is similar to standard pellets in appearance but of higher typical durability. This device may find wide application in speciality feed production such as horse feeds.

Odor Control

Given current trends in dust emission control, the next step may be odor control for feed mills. This is a huge issue with swine production units and is becoming so with many other types of production units. Technology is presently available to reduce odor emission to near zero. However, it is costly and adds no value to the product. If adopted, it will be done only under regulatory enforcement.

Hygienic Treatment of Finished Feeds

Feed manufacturers are under pressure to provide clean, pathogen-free feeds that do not contribute to the health problems of the target animal. Many manufacturers are investigating or initiating Hazard Analysis and Critical Control Point programs to accomplish that goal. Present technology in the form of hydrothermal treatment (i.e., expanders, compactors, UPCs, etc.) can do a great deal; however, it will take a great deal in the form of facility design and management skill to be successful.

In-Line Pellet Moisture and Quality Control

Several prototype and production models of in-line pellet moisture monitors are available. For successful adoption, these monitors must be given feedback control capabilities to be able to control residence time and air flow in coolers. Both temperatures and moisture content are important to shelf-life and must be controlled.

In addition to controlling coolers and dryers, in-line moisture monitoring and control is being adopted in the pre-pellet area. The concept is to refine management of feed moisture to improve pellet quality and production rate while controlling moisture shrink in the final product. At least two suppliers are actively promoting this technology in the feed industry.

Liquid Ingredients

Several major vitamin and animal drug manufacturers are pursuing developments of liquid forms of their products. In addition, several enzyme preparations, designed to improve nutrient availability, are entering the market (e.g., phytase). Often these have a use level of a few ounces per ton. There are several reasons for this development, not the least of which is a reduced cost of manufacturing and improved nutrition that may be shared with the final user. With the development of mass flow metres and other highly reliable liquid proportioning systems, accurate application rates are feasible down to less than one-half kilogram per ton. When this is coupled with the fact that liquids, when applied properly, do not segregate, it is easy to understand the interest.

Feed cost represents the major item in the cost of animal production. Without doubt, efforts will continue to refine feed manufacturing techniques to reduce the cost of feed and to increase its value to the target animal. The possibilities for improvements in feed manufacturing are endless; however, the cost of each innovation must be carefully weighed against demonstrated improvements in animal performance. In some cases, changes in feed manufacturing technology will be dictated, not by animal response, but by other motivations such as regulatory guidelines or health concerns. A case in point is the use of hydrothermal processes, such as pelleting, extrusion, or roasting, to reduce the microbiological load in the feed.

Optimising Liveability of free-range Layers

Optimising liveability of free-range layers is becoming an important welfare and production issue. A recent epidemiological study found the most common causes of death for free-range layers are reproductive tract lesions and cannibalism. Other causes of death, including infestation with parasites, attack by predators, mechanical rupture of blood vessels (heart and liver) and chronic disease caused by stress and malnutrition, together account for a very small percentage of deaths. For free-range farmers, the following suggestions will help combat these causes of death.

Rearing Viable Free Range Pullets

Rearing of healthy free range pullets requires quality chicks, strict biosecurity measures and a thorough vaccination regime. Vaccine titres should be monitored throughout the rearing phase and through production to avoid incorrect vaccination methods and/or degradation of vaccine efficacy from impacting on the effectiveness of the vaccination regime. Ensuring bird weight is equal to or greater than breed standard is also essential to successful rearing.

Management Approaches to Reduce Free-range Layer Deaths

A vaccinated layer with body weight equal to the breed standard provides the basis for a healthy flock. Ensuring biosecurity is strictly adhered to and providing appropriate nutrition, protection from predators, early intervention and treatment of sick birds, adequate nests/perches and floor egg management should minimise deaths.

Prevention of Bacterial Infections in Free Range Layers

Bacterial infections affect birds on the ground more than those in cages, which impacts on bird welfare and adds to production costs. Ways to combat bacterial infections include strict biosecurity (including all in all out flocks), medication, separating birds that are infected, reduced stress (malnutrition, overcrowding and dirty conditions), vermin control and clean waterers and feeders.

Importance of Correct Disease Diagnosis

If the symptoms of diseases are similar (for example, fowl cholera and spotty liver) there is an increased chance of farmer misdiagnosis, particularly if the farmer is more familiar with one of the diseases. Increased farmer education and improved use and availability of poultry veterinarians may result in an early and correct diagnosis decreasing the impact on bird welfare and production.

Alternative Poultry Production Systems

In Australia there are three main alternative production systems adopted by poultry producers. These include free range and organic, for both the production of meat and eggs and barn, for egg production. Such production systems provide the consumer with a choice over the methods by which their food is produced and includes issues associated with taste, welfare and environmental sustainability. Alternative poultry production systems are seen as complementary to conventional poultry production, however there have been some issues where both conventional and alternative production systems have been unfairly criticised by one another. Ultimately the goal for the entire industry is one of cohesion between production systems so as to promote consumer confidence in poultry as a whole.

Free Range

Free range is the poultry housing system most actively supported by welfare groups for egg production. Currently there are 1.69 million free range hens in Australia (11% of total laying hens in Australia) with an estimated grocery market share of 14.5%. Free range eggs are worth 23% of the value of the Australian egg industry, which is more than $71 million a year. Free Range poultry system for commercial production of eggs or chicken meat is considered an environmentally sustainable land use practice. This is achieved by limiting the stocking densities of birds outside the shed.

Under the Model Code of Practice for the Welfare of Animals (domestic fowl), free range allows poultry the maximum access to its natural behavioural instincts. Free Range hens are free to roam during daylight hours but are secured at night as protection from inclement weather and predators. Feed and potable water points are available within the shed housing which is in accordance with Bio-security regulations. Beak trimming is acceptable when carried out by accredited personnel only when all other treatments have been exhausted and must be initiated under veterinary instructions. In the free range system beak trimming at day old is not adopted. Human contact within the free range system is vital as constant interaction is necessary during hand egg collection, cleaning and monitoring stock.

Limited stocking density per shed is required to allow the hens' freedom of movement based on the Farm Animal Welfare Code 'five freedoms'. These are:

1. Freedom from hunger and thirst
2. Freedom from discomfort
3. Freedom from pain, injury and disease
4. Freedom to express normal behaviour
5. Freedom from fear and distress.

Free range chicken meat accounts for approximately 4% of chicken produced, with about half of it also being organic. Free range meat chickens are produced using similar management, housing and feeding practices as conventional meat chickens. The major differences are that free range chickens are allowed access to an outside run for part of each day (at least post the brooding period), have lower target stocking densities and some health issues are managed differently. Antibiotics can be used to treat sick birds but, once treated, the meat from these birds cannot be sold as free range. The main certifier of free range chicken meat in Australia is Free Range Egg and Poultry Australia Ltd (FREPA). The standards that free range meat chickens must comply with to be certified by FREPA. The major suppliers of free range chickens are two of the three largest companies operating in the commercial chicken meat industry.

Barn

The Barn poultry layer system is sandwiched between the cage system and free range system. Barn hens are housed on deep litter in secure and weatherproof sheds. While they are provided with ample water, feed, ventilation, space and nesting boxes in accordance with the Model Code of Practice for the Welfare of Animals (domestic farm), they do not range out of doors. Limited stocking density allows room for the opportunity to express their full range of natural behaviours. They have the freedom to flap their wings, stretch, fly and dust bathe. This allows hens to move around and explore their environment. The extra space also allows submissive birds to avoid dominant birds if they need to, which can reduce aggressive pecking and improve welfare. Hens have access to a private, enclosed area for laying, which is a priority for hens. Beak trimming is generally allowed with most birds' beak trimmed at day old to limit feather pecking and cannibalism. Barn sheds are usually fully automated with central nesting boxes over conveyor belts which carry the eggs out of the shed into a packing area, thus avoiding human contact.

Certified Organic

Only free range hens can be used to produce certified organic eggs and meat and must be grown without the use of artificial colours and synthetic chemicals, according to the national standards. Birds are fed certified organic feed grown from soil that is certified as organic and does not contain pesticides or inorganic fertilizers. Under the regulations, no antibiotic medication is to be used for treatment of organic poultry. Organic standards for egg production specify the types of feed, accommodation, and living conditions which are suitable for organic laying hens. While healthy hens require shelter for security and protection against the elements, they also benefit from the freedom to scratch and forage naturally outdoors. A clean, pollution free range provides a suitably varied natural diet, one that is rich in essential vitamins, minerals, and trace elements. Certified organic meat chickens have two additional requirements as well as having to be free range:

Feed must be predominantly from certified organic ingredients.

Birds cannot be treated with routine vaccination. There are exceptions, such as where treatment is required by law or disease cannot be controlled with organic management practices.

Certified organic chicken meat bears a certification logo from an approved organisation, and producers must comply with The National Standard for Organic and Bio-Dynamic Produce. Note that at present chicken meat can be described as 'organic' without being certified by an organic association. Therefore it is important to look for a relevant certification.

General tips for Small Scale Poultry Production

The following tips on aspects of small scale or backyard poultry should only be used as a guide. Poultry Hub accepts no responsibility for persons who use this information to grow or prepare poultry and poultry related products.

Raising Backyard Chickens

Choosing Backyard Chicken Breeds – Many chicken breeds are perfectly suited for backyard flocks. Choosing the right breed for your backyard flock solely depends on the poultry keeper's needs. Chicken breeds can be divided into two basic categories: birds for meat and birds for eggs. Also certain chicken breeds can boast milder temperaments, which are typically better suited for families with small children who would interact with the birds.

Chicken Coop Basics – Chicken coops must provide shelter and protection for the animals. Coops must be big enough for the chickens to live comfortably by allowing them room to flap their wings, walk about, and build comfortable nests. Chicken coops must also be hygienic—featuring changeable flooring (like sawdust, woodchips or sand), room for humans to clean, and ventilation. Chicken coops must also provide warmth through insulation, heat lamps, or extra bedding and nesting material.

Catch, Check and Dispose of Spent Birds

Culling is removal of individual birds from the flock and in most cases, humanely killing them. Culling is an essential management practice. It is part of maintaining the health of the flock and high egg production. Sick or diseased birds that can't be treated should be killed as part of preserving the health of the rest of the flock. It is also pointless keeping non-laying bird unless it is a pet. If the bird is not sick, but still needs to be culled, it can be used for food consumption. Modern hybrid layers are not really suitable as table birds as they are light and small framed, with little meat on their bones. Other breeds are more suited for the table. If you want to preserve high egg production of your flock, you will choose to cull all birds that have completed two complete laying cycles in an all-in all-out system. In a culling operation, you should act in a quiet, deliberate, confident and humane manner.

Catch and Hold a Bird

The following points are useful in catching and holding birds:

- Quiet birds are the easiest to catch.
- Pen the bird in a confined space first.
- Birds are quiet and easy to manage in low light conditions (eg dusk).
- A catching hook may be used to hook the animal's leg.
- Approach the bird quietly and make slow deliberate movements that block escape.
- Grasp the wings to the body by moving your hands down over the top of the bird.
- Then grasp the legs in one hand while continuing to hold a wing to the body with the other.
- Support the bird's body with the hand holding the legs and when it has settled, the bird should be easy to inspect as required.

Check Health and Laying Condition

Poultry Agskills – A Practical Guide to Farm Skills

While holding the bird securely, check its general health and condition and its laying status:

1. A good layer will have a large, smooth, moist, pinky-white vent.
2. The two small bones at the sides of the vent are called the pubic bones.
3. They should be flexible and wide apart, with at least two finger widths between them (one finger width = 2cm).
4. The abdomen should be deep, soft, and pliable without an accumulation of body fat.

5. The depth of the abdomen is measured between the tip of the keel or breast bone and the pubic bones.
6. Laying hens should have a depth of three or four finger widths (6 to 8 cm).
7. The non-layer will usually have a smaller body with a shallow, firm abdomen.
8. Pullets and non- laying hens have a depth of about two finger widths (4cm) between the pubic bones and keel.
9. The pubic bones are usually stiff and close together when the hen is not laying the distance between the pubic bones is one finger width or less.
10. The vent of a non-layer is usually small, puckered, and round.

Humanely Kill a Bird

- If right handed, grasp the bird's shanks in the left hand with the breast of the bird facing away from you.
- Hold the bird's head in the right hand, placing your thumb over the bird's head at the point where the neck joins the head. Wrap your four fingers around the throat area of the lower head, with the little finger against the bird's bottom beak.
- With the bird's breast resting across your right thigh, twist the beak upwards with the fingers, while holding the back of the head down with the V formed between the thumb and index finger. The bird's head should now be at right angles to the neck.
- While holding the head in this angled position, quickly and forcefully pull the head downwards in a short, sharp movement, all the while holding the legs firmly in the other hand.
- This extension will dislocate the bird's neck immediately behind the skull.
- When performed correctly, you will feel the dislocation of the neck and it will create a cavity between the head and neck. This severs the spinal cord and jugular vein instantaneously. Killing is quick, clean and humane.
- Continue the slight downward pressure until there is a 1–2 cm gap between the head and the top of the broken neck.
- If the head is kept hanging down the blood will drain into this cavity.

- Do not pull the head too far or it will be severed from the body.
- Wing flapping will normally occur. This is simply a muscular reaction, although the bird is dead. The flapping should stop within approximately 30 seconds.

Dispose of the Carcase

1. Birds that are sick before being killed should not be eaten.
2. Dispose of these carcases by immediate burial, burning or by taking to a vet.
3. If burying a carcase, ensure it is buried deep enough to prevent retrieval by dogs or foxes.
4. If burning the carcase, use a closed incinerator to reduce the airborne spread of disease.

Egg Cleaning and Salmonella

In the Australian context, it is generally accepted that we do NOT have the types of Salmonella that can transmit from the hen to the egg. This type of Salmonella does exist elsewhere in the world – Therefore people should be very careful about where they order sunny side up eggs or have real egg mayonnaise.

Faecal contamination on eggs is a source of Salmonella and this concern is amplified with cracked eggs and even worse, cracked and dirty eggs. Therefore, people should have concerns about egg washing. If washing is not done correctly (e.g. if the washing is done at the wrong temperature), there is a possibility of the natural barriers of the egg being compromised. Hence, poor quality washing can actually be more of a risk, causing the outside bacteria to be sucked inwards into the egg. Good, well managed egg washing is beneficial, conversly poorly done washing is more likely to cause a problem than solve the problem. Finally, Salmonella is a very good survivor in the environment, hence any compromise of the natural barriers of the egg can provide an opportunity for Salmonella. Therefore one should seek fresh eggs with no obvious shell damage and prefer washing to be done in a professional and controlled manner. If in doubt, simply go for the well cooked egg (hard boiled eggs)!

Key Points for Good Egg Washing

Very dirty eggs should first be dry scraped to remove the bulk of the material. Use warm detergent solution (43-52°C) – important that this temperature is always higher than the egg temperature (around 15°C higher but not greater than 50°C) and change wash solution if clearly dirty.

Overheating Eggs

As a home consumer, it would be unlikely that you are going to temperature abuse (overheat) your backyard eggs or store them for a prolonged period. It is this temperature abuse and prolonged storage, when combined with subsequent inappropriate food handling that results in food-borne disease. In Australia, the types of Salmonella present in our food production animals have a large infectious dose. You really need to consume large numbers or the organism to acquire the disease (this is why temperature abuse is such a critical factor in food-borne disease outbreaks – the Salmonella needs time and right conditions to multiply). Poor egg washing can result in a compromise of the natural egg barriers and entry of faecal Salmonella into the egg. In conclusion, given that there is no temperature abuse or prolonged storage, and good kitchen hygiene, it is unlikely that poor egg washing will have an impact on you.

Identify Stock

There are two main ways of identifying poultry. The first is to identify the physical differences between the sexes, which is particularly important in culling and separating female layers from the males. Males may be selected for further breeding or culled. Identifying individuals can be done in a number of ways. Birds may be wing-tagged or have a numbered or coloured legband attached. Birds used in a breeding program are often identified by toe punching, in which a hole is made in the web of the feet in one of 16 different combinations. If an individual needs to be separated from the flock for mating, breeding, medicating, quarantining or to break broodiness, they can be caught, identified and housed separately.

To Identify and Draft Poultry

When you purchase day-old chicks from a commercial hatchery, they have been expertly identified either by vent-sexing, or by wingfeather examination. The wing feathers of a female chick are thick and well-developed, while the male has smaller, thinner wing feathers. In some cross-breeds, the day-old progeny can be sexed by colour. There is a genetic link between colour and sex. For example, in across between a Rhode Island Red male and a White Wyandotte female the male day-old chicks are white and the females are reddish. It is difficult for a backyard breeder to identify the sex of birds until they are four to five weeks of age or even older. Males are faster growing and by six weeks of age have bolder heads, longer necks and show reddened wattles and comb. Feathers on the back (the saddle) of females are blunt, dense and wide. Feathers on the back (the saddle) of the male are pointed, less dense and distinctly tapered.

Feed Evaluation

Feed evaluation is the testing of feed quality, providing information on the composition of feed or feed ingredients as well as their suitability for poultry. Poultry feed is made up of many ingredients, which are broadly grouped into providers of energy (fats, oils and carbohydrates), protein (amino acids), vitamins, minerals and product quality enhancement. Typically, cereals such as wheat, barley, sorghum and maize will provide energy while soybeans, lupins, canola and peanuts provide protein. These ingredients are then combined in such a way as to provide the energy, protein, vitamin and mineral requirements for poultry through the process of feed formulation. In order to know what amounts of these ingredients should be included in the diet, the ingredients are first evaluated, to see what nutrients they contain in what quantities. After the diet has been prepared, it may also be necessary to evaluate the complete product, to determine its suitability for the class of poultry that will be fed (such as egg layers, meat chickens or breeders). Feed evaluation is a key process in the poultry industry. Feed ingredients need to be tested in order to formulate the complete diet, and diets have to be evaluated to determine their suitability for poultry. Evaluation provides different types of information, as required by nutritionists and farmers. In general, the range of tests that can now be performed is wide and it is now possible to obtain results rapidly.

Measures of Feed Quality

Feeds and feed ingredients can be evaluated physically as well as chemically. The physical evaluation of feed mostly provides preliminary information on the quality of the material. It involves assessing physical qualities such as weight, colour, smell and whether the material has suffered from any contamination by other materials. Chemically, feed is made up of water and dry matter. The dry matter contains organic and inorganic compounds. The organic part of feed is made of mainly carbohydrates, proteins, vitamins and fats and oils. The inorganic part is made of mineral elements, also known as ash. Feed or feed ingredients can be analysed to provide values of each of these components. Apart from obtaining values of chemical composition, the extent of utilisation of these components by the bird, termed nutritive value, is also measured.

How Feed Quality is Measured

Feed quality is measured by chemically breaking up the food into the components mentioned above. In the industry, it is sometimes necessary to break down these large components into smaller analytical fractions. Thus, values of starch and the non-starch component (called fibre) of carbohydrates may be provided. Proteins are made of amino acids, 10 of which must be present in poultry diets, so their amounts should be indicated during feed evaluation.

In the past, feed evaluation was a cumbersome process, requiring days to complete. However, newer equipment and procedures have been developed, which enable the rapid evaluation of most materials. For example, energy content is rapidly determined on bomb calorimetres, linked to computers, to eliminate the need for rigorous calculations. Protein is rapidly determined on Leco® machines, which eliminate time-consuming digestion of feed with strong acids and reaction of material with acids and bases. Near-infrared reflectance spectroscopy is one of the latest techniques by which feed can be evaluated with the most minimal preparation of the sample.

Of greater importance in feed evaluation is the response of poultry to particular feeds. This is regarded as the real nutritive value of the feed, and must be measured as part of feed evaluation. Nutritive value does not necessarily entail animal growth or egg production. It gives information on how much of each of the fractions in feed, i.e. starch or energy, protein (amino acids), fats, minerals or vitamins, was used by the bird. When feed is given to poultry, they are able to break down only a fraction of the feed and absorb it into the body for growth and egg production. The rest is voided in faeces and urine, which are excreted together by poultry. The amount of nutrients retained by the bird is an indication of the nutritive value of the feed.

Importance of Feed Evaluation

Feed evaluation is important because ingredients that belong to the same class contain different nutrients; for example, maize provides more energy than wheat while soybeans contain more proteins than lupins. The same ingredient varies from one supplier to the other, and between years. In drought years, cereals fill poorly and are therefore lower in quality. Most importantly, if feeds are not evaluated, it is not possible to tell if the material will be suitable for feeding poultry. Feeding standards have already been set for different types of poultry, so the requirements for different nutrients must be met precisely. It is possible, with the current state of knowledge, to predict poultry growth or egg production by modelling feed quality, type of housing, class of poultry and duration of feeding. The central key issue in these models is feed quality, which can only be obtained through feed evaluation.

Impacts of HPAI on Rural Livelihoods: Conceptual and Analytical Frameworks

Seventy percent of the world's rural poor depend on livestock as a component of their livelihoods (LID, 1999; FAO, 2002), and a vast majority of those keep poultry (Sonaiya et al., 1999; Epprecht et al., 2007). Village poultry production, which is also known as backyard extensive poultry production, is a common phenomenon in many developing countries. This mode of poultry production is characterised by ownership of up to a hundred birds; generally reared free-range, with minimum or low inputs and zero to minimal biosecurity (Guèye, 2000; Kondombo et. al., 2003; Muchadeyi et. al., 2005; Abubakar et. al., 2007). In fact among the rural poor, poultry is found to be a crucial livelihoods asset for the poorest segments, such as those households that are in the first income quintile (Maltsouglou and Rapsomanikis, 2005; Roland-Holst et al., 2007). Poultry production by rural poor households contributes to several livelihoods indicators, including (but not limited to) income, nutrition, food security, savings, insurance and gender equality (Campbell and Tretcher, 1982; Watts, 1983; Sonaiya, 1990; Sonaiya and Olori, 1990; Chitukuro and Foster, 1997; Kitalyi, 1998; Kushi et al., 1998; Sonaiya et. al., 1999; Guèye, 2000; Sonaiya and Swan, 2004; Muchadeyi, 2005; Alabi et al., 2006; Abubakar, 2007; Guèye, 2007a; 2007b; Sonaiya, 2007; Smucker and Wisner, 2008). Furthermore poultry production constitutes a quick and high return investment opportunity (Epprecht et al., 2007; Sonaiya, 2007) for improving any one or all of these livelihoods indicators. Moreover, poultry production is often recognised as an entry point into livestock production (Alabi et al., 2006; Guèye, 2007a), which is associated with breaking out of poverty traps.

The roles of poultry in income and food security are straightforward to characterise: rural poor households generate cash income through the sales of those poultry and poultry products (e.g., eggs, meat, feathers) which they do not consume themselves. With the income generated through these sales they are able to buy other types of food and/or inputs to produce other types of food (e.g., seeds to produce grains). Contribution of poultry to other livelihoods indicators, namely savings, insurance, nutrition and gender equality, however, requires further elaboration. In rural areas where credit markets are missing, similarly to other large livestock, poultry functions as 'insurance' to hedge against shocks and stresses (Binswanger and McIntire 1987; Rosenzweig and Wolpin 1993; Fafchamps, et al., 1998). Often poultry functions as a 'savings account', which can be tapped into fairly quickly to meet household needs such as school fees, costs of weddings and funerals (Obi et al., 2008). Poultry represents a store of value which appreciates very quickly with time, as demonstrated by the high productivity parameters estimated for village/backyard extensive poultry across developing countries (Sonaiya, 1990; Kitalyi, 1998; Sonaiya and Swan, 2004).

Poultry also contributes to household nutrition, as many rural poor households rely on their own poultry production to supply the majority of their animal source food. Poultry provides not only protein but also highly-bioavailable essential micronutrients, such as iron, Vitamin A and zinc, which are crucial especially for child nutrition and health (Iannotti et al., 2008). Chronic malnutrition and micronutrient deficiencies are very high in developing countries (Quinn et. al, 1990; Callens and Phiri, 1998) and hence poultry is particularly important for the improvement of this livelihoods indicator. In fact several ongoing projects in developing countries, especially in African countries, are encouraging livestock production, particularly poultry production as a nutritional intervention (Narrod, 2006).

Finally, village poultry or backyard extensive poultry production is important for empowering women and promoting gender equality in developing countries. In most developing countries village poultry or backyard extensive poultry are mainly owned and/or managed by women, providing them (as well as those they look after including children, elderly and the invalid) with livelihoods (nutrition, food security, income, wealth, and consequently empowerment) and hence contributing to gender equality. In fact World Hunger Project lists poultry as one of the new income earning activities in which women are increasingly being involved (The World Hunger Project, 2005). In Africa for example, women are widely recognised to be the main poultry keepers, constituting over seventy percent of all poultry ownership (Alder, 1996; Guèye, 1998; 2000; Kitalyi, 1998; Sonaiya, 2007). Household level studies conducted in Africa revealed that women earned significant incomes from poultry sales, even after accounting for household consumption (Chitukuro and Foster, 1997; Kushi et al., 1998, Alabi et. al., 2006). Similarly, in Indonesia, amongst all livestock types, women are most likely to own poultry, and within households women are as engaged as men (if not more) in all aspects of poultry rearing (Asare-Marfo and Birol, forthcoming).

Recent outbreaks of Highly Pathogenic Avian Influenza (HPAI) in Africa and Asia are therefore expected to have detrimental impacts on the rural poor households' livelihoods indicators, especially on those pertaining to income, food and nutrition security, and gender equality. A review of the literature on economic impacts of transboundary animal diseases reveals that rural poor households suffer greater impact when poultry disease outbreak occurs (FAO, 2002; Otte et. al., 2004; 2006; FAO, 2008).

Rural poor are especially vulnerable to HPAI or other poultry disease outbreaks due several reasons. First, rural poor households keep few numbers of birds and the poorest households obtain highest share of their income from poultry (Roland-Holst et. al., 2007). Second, rural poor have limited or no access to infrastructure and inputs (e.g., vaccination, biosecurity, balanced feed) which are necessary for the minimisation of HPAI contamination risk. Third, an outbreak may wipe out the entirety of a rural household's flock, thereby removing their entitlement for compensation.

Finally, even though information on HPAI and education/training for biosecure production may be available for some rural poultry producers, poorest segments and women poultry producers may not have access to these. Most of the HPAI research to date has been epidemiological, focusing on the prevention of the infection both among poultry, and from poultry to other livestock and humans. There have been several economic studies which focused on the impacts of HPAI and other poultry diseases on production losses and hence on income, with little or no attention paid to the impacts of HPAI on livelihoods strategies and outcomes other than income (Hall, 2002; Anderson, 2003; Kinung'hi et. al., 2004; FAO, 2006; Akpabio et al., 2007; Burgos and Burgos, 2007; You and Diao, 2007). Although there have been some theoretical studies (e.g., Beach et al., 2007a; 2007b) limited research efforts have been directed towards investigating the efficient and effective ex ante and ex post disease control policies/programmes to reduce HPAI risks and to mitigate the impacts of HPAI on affected households. Therefore, there is currently very scant evidence-based research on the impacts of HPAI outbreaks and threats on the livelihoods of rural poor and the cost effective strategies to minimise HPAI risks.

One of the aims of the Pro-Poor HPAI Risk Reduction Strategies project funded is to fill in these gaps by conducting in-depth research on the impacts of the HPAI outbreaks and threats on rural livelihoods, specifically to measure the rural households' vulnerability to this livestock disease and its associated repercussions (e.g., food and nutrition insecurity and deepened gender inequality). The key rural livelihoods and vulnerability related research questions that will be investigated in this project are:

i. How do HPAI outbreaks and threats affect rural households' assets; livelihood strategies that can be achieved with these assets and the resulting livelihoods outcomes?
ii. To what direction and extent do household assets, livelihood strategies and outcomes change in response to various risk minimisation strategies and changes in disease control policies?

The aim of this paper is to provide a methodological framework that will enable the collection and analysis of primary data to measure the impacts of HPAI on assets, livelihood strategies and the various indicators of rural livelihoods, as described above. To this end, the Sustainable Livelihood Framework (SLF) and complementary qualitative and quantitative methods are proposed to answer these two broad research questions. More specifically, the paper describes how the SLF; a combination of data methodologies and sources can be applied to study the impact of HPAI on rural livelihoods, and to evaluate strategies to minimise HPAI disease risk. Of specific focus in this framework is the identification of those quantitative indicators which can detect and measure the effects of HPAI on livelihoods, and can therefore inform the development of cost effective risk minimisation strategies and policies. The rest of this paper unfolds as follows: The SLF, as the underpinning conceptual framework, is explained in the next section. This is followed by the description of the conceptual framework adopted to study the key research questions; the resulting main hypotheses and review of those studies which tested similar hypotheses. The analytical framework proposed to test these hypotheses and the data requirements are explained in section 4. Section 5 discusses the application of these proposed conceptual and analytical frameworks in selected case study countries.

The Sustainable Livelihoods Framework

Description of the Sustainable Livelihoods Framework

Livelihoods research is linked to the 'New Poverty Agenda' of the 1990s, which recognised the deficiencies of a solely income-based measurement of poverty, and adopted a multi-dimensional view of poverty including indicators to measure improvements or shortcomings pertaining to health, education and environment (Prowse, 2008). The livelihood concept is based on the premise that a rural household has access to (or has an endowment of) a minimum amount of resource base (i.e., capital or assets), which can be utilised to fashion out a set of livelihood strategies (e.g., crop farming, livestock rearing, off farm employment, etc.) to improve household welfare (Chambers and Conway, 1992).

A household's livelihood is sustainable if it can cope with and recover from shocks (such as civil conflict or emergence of new human, crop or livestock diseases) and stresses (e.g. recurrent adverse weather and seasonality) seasonality); maintain or enhance its capabilities and assets, while not undermining the natural resource base. The SLF proposed by Chambers and Conway (1992) was expanded by DFID (2000) and became a popular tool within development studies, as well as for development policy and planning (Prowse, 2008).

The popularity of this approach stems from the numerous benefits it has brought to development research and policy, such as the opportunity to conduct inter-disciplinary and cross-sectoral analysis, and linking both macro and micro levels. The DFID SLF is presented. As it can be seen the framework consists of four elements: a vulnerability context, an asset pentagon, transforming structures and processes, and livelihood strategies. The framework shows that these four inter-related elements contribute to livelihood outcomes (Ashley and Carney, 1999; Carney et al, 1999; DFID, 2000).

According to the SLF presented above, an assessment of the impacts of a shock and/or stress on livelihoods should begin with an analysis of assets i.e., a review of what people have (Carney et al., 1999). Among the assets available to an agricultural household are natural capital (N), physical capital (P), financial capital (F), human capital (H), and social capital (S), though this list is not exhaustive (Scoones, 1998). Not depicted, but also considered as important assets are political capital (Tofique, 2001) and location capital (Jansen et al., 2005). The quantity and quality of these assets and access to hereof are influenced by the vulnerability context, including trends, shocks and stresses.

Some examples of shocks are natural disasters, civil wars and HPAI outbreaks, whereas trends are expected disturbances such as hungry seasons and endemic livestock or human diseases. The SLF depicts a set of transforming structures and processes, such as policies and institutions, which influence livelihood strategies. Based on the various interactions livelihood between assets and transforming structures and processes, the SLF delineates a set of livelihood outcomes, which could be a combination or one of these: more income; reduced vulnerability to trends, seasonality, shocks and threats; improved food security; reduced malnutrition and a more sustainable use of the natural resource base.

Households are viewed as being sustainable if they can cope with trends, shocks and seasonality without compromising their future ability to survive these. In addition to the DFID SLF, there are a number of SLFs that also take an asset-based and vulnerability approach to the analysis of rural livelihoods. These are CARE, Oxfam and UNDP livelihoods approaches, all of which are based on the livelihoods definition of Chambers and Conway (1992). Similarly to the DFID SLF, all of these SLFs stress the need to understand micro and macro (and policy) links.

All of these SLFs identify the contexts, conditions and trends within which a rural household combines its assets to generate bundles of livelihood strategies. These frameworks indicate that the policy context of a rural economy; the trend of socio-economic factors (such as population and macro-economic conditions) and other exogenous factors (such as agro-ecological conditions) determine the type of assets available to the rural households. These SFLs however differ in a few aspects, such as the definition of assets and analysis procedures (Carney et al., 1999).

Adaptation of the Sustainable Livelihoods Framework to HPAI Shocks and Threats

Project focuses specifically on those rural poor households that are engaged in the production of poultry, either as one of the several livelihoods strategies (i.e., village extensive/backyard extensive poultry keepers) or as the main livelihoods strategy (i.e., backyard intensive or small scale (semi) commercial poultry producers/farmers).

Hence, adaptation of the SLF to this context requires redefinition of the assets that are available to these households and the asset-modifying factors (i.e., transforming structures and processes depicted). At a household scale, a poultry-producing household (depending on the agroecology of the environment) is expected to have natural capital (land: inherited or acquired; water: stream, borehole, treated water; soil: quality and fertility; genetic resources in the form of disease resistant breed of chicken, etc), physical capital (numbers and types of livestock, production equipment and technologies, basic infrastructure (e.g. building), machinery, transportation), social capital (social networks, social relations, membership of national or village level poultry associations, etc), human capital (number of household members, their gender and age compositions, skills, knowledge (indigenous/local or formal through extension training), (in)formal education, good health, ability to work, household size and demographics), and financial capital (cash, credit/debit, savings).

In addition to these traditional definitions of capital, there are other capital/assets such as information, location and political capital, which might have impacts on the livelihood strategies and resultant outcomes.

Information capital might include the type, amount and source of information a poultry-producing household has about biosecurity, HPAI or other poultry disease risks and disease prevention This capital would determine households' ability to manage poultry health and have impacts on their attitudes towards risk.

Location capital constitute all location specific factors, such as access to infrastructure (e.g., roads) and markets (Jansen et al., 2005), as well as being located in areas where HPAI risks could be higher (e.g., near wetlands or borders where there might be some illegal trade of poultry and poultry products).

Political capital describes farmers' social standing with the administrators (such as the local government council, political office holders, local elites and head of a community, to name a few) which might have implications on farmers' access to public sector interventions regarding disease control (e.g., subsidized inputs such as drugs, disease-resistant breeds, vaccinations, funds (e.g., subsidies or credit) for biosecurity investment or compensation in case flocks are culled) (Tofique, 2001). Together these assets constitute a stock of resources used to generate well-being (Moser, 1998; Rakodi, 1999).

The productivity of these assets, and hence their contribution to the overall livelihood outcomes often depends on their complementarity. In the case of poultry assets, for example, poultry may have different implications on livelihood outcomes depending on the distance to the live bird markets and/or access to transportation as well as means of transportation. Furthermore the institutional context (e.g., political, regulatory and legal frameworks), in which households operate affect management and productivity of assets and the type and success of livelihood strategies undertaken.

For example, strict disease control policies (e.g., hose ensuring high levels of biosecurity) could render village level poultry keeping an obsolete livelihoods strategy, but if government subsidies and training and extension programmes are available, households could invest in biosecure backyard poultry production.

A household's management of its asset portfolio constitutes household behaviour or livelihoods strategies, which refer to the ways the household combines its assets to generate well being (Jansen et al., 2005).

For example, households may combine their poultry assets with their knowledge of healthy poultry management to produce safe and nutritious poultry meat and eggs which may result in livelihoods outcomes of improved household nutrition or when combined with market infrastructure and transportation assets, these could result in the livelihoods outcome of poultry income. The SLF framework will enable understanding of the impact of the vulnerability context, which in this case is HPAI shocks and threats, on the other three components of the framework:

i. Livelihood assets: Changes in a household's asset portfolio, including, for example, changes in poultry assets (in quantity and value); changes in human capital in the form of information and education pertaining to poultry production; changes in poultry infrastructure (e.g., coops) to improve biosecurity.
ii. Transforming structures and processes: Changes in institutions, such as policies for disease control, surveillance, and changes in markets (e.g., demand, prices etc.).
iii. Livelihood strategies: Ex post HPAI coping strategies, such as substituting poultry meat and eggs for other less nutritious foodstuff (e.g., vegetables), and ex ante and ex post risk management strategies, such as diversifying into other livelihoods strategies, such as crop farming or other small livestock rearing.

Impacts of HPAI on all of these components of livelihoods are expected to affect various livelihoods outcomes including poultry income, wealth (savings/insurance), food and nutrition security, and gender equality. The interrelationships between these three components of the SLF in the context of HPAI shocks and threats and the impacts on livelihoods outcomes are explained in greater detail below, where the resultant hypotheses are also stated.

A Conceptual Framework for Investigating the Impacts of HPAI Shock and Stress

The aim of this paper is to develop a framework to assess the livelihood impacts of HPAI outbreaks (shocks) and threats (stresses) and disease control strategies and policies on household livelihoods indicators.

To this end based on the previous work by Reardon and Vosti (1995), Orr and Mwale (2001), Jansen et al. (2005), Sesabo and Tol (2005), Iiyama (2006) and Roland-Holst et al. (2007), the basic SLF presented above is augmented to conceptualise the analysis of household response to HPAI outbreaks and threats. As explained in the generic SLF above, vulnerability contexts, such as the HPAI shocks and stresses affect household's livelihoods asset portfolio, which includes physical capital such as the poultry asset base.

The impact of HPAI on poultry asset base in turn brings about coping strategies and mechanisms, which have implications for household behaviour or livelihood strategies (mainly those pertaining to poultry keeping/production) and hence on the overall household livelihood outcomes (income, food and nutrition security).

Household livelihood outcomes in turn determine the level of future vulnerability of the households to various shocks and stresses, including future outbreaks/threats of HPAI.

Transforming structures and processes, i.e., the institutional context, which in this case comprises policies pertaining to the control and management of HPAI (surveillance, depopulation, compensation, biosecurity laws etc), are influenced by the vulnerability context, i.e., HPAI disease situation, household livelihood strategies (e.g., poultry management, level of biosecurity), as well as household outcomes (e.g., income from poultry). At the same time, institutional context has implications on the extent and impact of HPAI outbreaks and threats on household asset base, household livelihood strategies and outcomes.

H1: HPAI shocks and stresses result in the loss of poultry assets and the extent of this loss depends on various household and community level as well as institutional factors.

HPAI outbreaks result in the loss of poultry either in volume (due to disease contamination and/or depopulation/culling) and/or in value, i.e., in the price (due to decreases in consumer demand as a result of outbreaks and/or scares).

Similarly, even if an HPAI outbreak may not affect the flock owned by a household, it may have detrimental impacts on the overall value of household's poultry assets. The magnitude and direction of the HPAI outbreak and/or threat on poultry asset value and/or volume are expected to be context specific, depending on various factors such as distance to the outbreak epicentre or live bird markets; the nature and efficacy of disease control policies; availability of substitute commodities/assets; efficiency, efficacy and reliability of risk communication, to name a few.

Almost all of the previous studies on the analysis of the impacts of shocks and stresses on rural livelihoods start by making an inventory of the assets the households have. Knowledge on type, quantity and quality of various asset types enables understanding of the complementarities and substitutions between each asset type and how they determine household livelihood strategies, as well as households' choices of coping strategies. Therefore detailed data on each asset type owned by the household should be recorded in the household surveys purposefully developed to study the impacts of HPAI on rural livelihoods.

Even though all of the assets described in section two are crucial for the formation of livelihood strategies and coping mechanisms against HPAI shocks and threats, specific attention should be paid to some of them. Detailed information on the number, type (species), variety and market prices, as well as flock dynamics of poultry should be collected.

In the rural livelihoods context poultry may count as natural capital (especially for the genetic resources present in rural poultry varieties) (Campbell et al., 2002), as well as physical capital (as a part of the productive assets, for example to provide meat and eggs for market sales and/or to meet the nutritional needs of household human capital). Moreover, similarly to other livestock, poultry is often considered as a major insurance asset, which households usually liquidate during short-term crisis (e.g. Campbell and Trechter, 1982). It is therefore important to verify for what purpose(s) rural households keep poultry so as to understand the impacts of HPAI induced loss in value and/or volume of poultry on household livelihoods strategies and their vulnerability to future shocks and threats.

Other assets that are thought to be important complements to poultry are human capital, location, social and political assets. Information should be collected on the gender and age composition of the household members, as well as on their education, training and experience, especially with regards to livestock in general and poultry in particular. Information on the ownership and management of poultry disaggregated by gender would shed light onto the gender issues in village poultry production system (Kitalyi, 1998; Guèye, 2005; Muchadeyi et. al., 2005) and will enable determination of the impact of HPAI on household livelihoods as well as on women's livelihoods. Moreover, data on the household size and demographics (especially dependency ratio) would provide information on the protein and micronutrient consumption patterns in households.

In addition, information on the current level of experience and education household members have, could be used to test if these assets can explain diversified livelihood patterns as found by previous studies (Iiyama, 2006).

Finally, information on poultry specific experience and education could help inform training interventions for biosecure poultry management. Information on other household (and community) specific assets such as location capital (e.g. access to infrastructure and public services, access to markets, remoteness, proximity to wetlands and border); social capital (village tradition, culture, religion), political capital (relationships with the local governments, NGOs) should also be collected.

All of these assets are expected to have impacts not only on poultry management, but also on livelihoods strategies and coping strategies (Pyle, 1992; Jansen et. al, 2005; Gilbert et. al., 2007) in the case of HPAI shocks and stresses. Finally information on other assets such as land, machinery, and other livestock should also be collected to be able to understand the various livelihoods strategies that might be available to the households before and after HPAI induced shocks and stresses.

H2: Poultry asset loss results in changes in the household behaviour/livelihood strategies.

Changes in the household behaviour or livelihoods strategies are classified into two categories: *ex post coping* decisions, following an outbreak, and *ex ante* or *ex post* risk management decisions pertaining to disease threats. *Ex post* coping decisions are short term asset disinvestments which may include segregation of species, rapid selling off of poultry or reduction in the consumption of poultry meat and eggs, whereas risk management decisions constitute longer term asset disinvestments, such as investment in biosecurity, training in biosecure poultry production, diversifying into other agricultural activities (e.g., crop production, other livestock rearing), or migration to name a few. Coping and risk management decisions are expected to be affected by several household specific characteristics (e.g., household asset portfolio, and in particular household's risk perceptions), as well as the institutional context such as the disease control policy (e.g., surveillance in live bird markets, compensation schemes, biosecurity requirements) and other public interventions (e.g., subsidies for strengthening of biosecurity, extension services of training in biosecure poultry production or other alternative livelihoods activities), as well as private sector services such as availability of credit.

Several studies to date have used the SLF or a similar asset-based approach to investigate *ex-post* coping and risk management strategies to shocks and *ex-ante* risk management strategies to stresses.

The nature of coping mechanisms adopted and their returns have been recognised as important channels for tracing the livelihood impacts of various types of stresses and shocks, such as drought, famine, civil war, macro-economic policy change as well as household level stressed and shocks such as illness, death, divorce and indebtedness, to name a few. These studies on coping mechanisms found that the initial asset endowment have the greatest impact on the coping strategies, with poor asset endowments resulting in low–return coping strategies and exposing poor households to increased vulnerability to shocks.

Moreover, the role of livelihood diversification as a risk management strategy (e.g., against droughts, loss of various necessary inputs such as land and water) as well as a *ex post* coping strategy has been investigated by various studies, many of which have also estimated the impact of diversification on livelihoods outcomes. Overall these studies have shown that rural poor do not specialise in one activity only, i.e. they diversify their livelihoods strategies, and the more diverse are the livelihoods systems, the less vulnerable they are to shocks and threats and the higher the livelihoods outcomes they yield (Ellis, 2000; Iiyama, 2006).

In those countries where HPAI outbreaks have occurred, an assessment of the impacts of HPAI outbreaks, i.e., shocks, on the livelihoods outcomes of the rural poor should focus on the investigation of the nature and extent of *ex post* coping strategies. To this end, both qualitative and quantitative household, community, regional and national level data should be collected on the exact sequence of the events resulting in the shock; resulting market conditions; asset endowments and livelihoods strategy portfolios prior to and following the shock (Corbett, 1988). Similar data, as well as information on the households' and communities' knowledge of HPAI and risk perceptions and attitudes should be collected to investigate the impact of HPAI threat, i.e., stress on household *ex ante* risk management behaviour.

H3: Changes in household livelihood strategies result in changes in livelihood outcomes.

HPAI outbreaks and threats are expected to have significant and negative short run and long run impacts on several measurable household livelihood outcomes of poultry producing households, such as household income from poultry, food and nutrition security, and gender equality, as explained in the introduction.

The long run impacts of HPAI, however, could also be positive, if households improve/invest in their poultry management practices (e.g., biosecurity levels, feed and other factors that affect poultry immune systems) as a result of increasing awareness and/or due to compensation and/or other public sector schemes.

The net long run impact of HPAI on household livelihoods outcomes could be either positive or negative depending on various factors, such as household specific characteristics (e.g., household assets), as well as the institutional context, such as the compensation schemes and/or availability of retraining in alternative livelihoods activities. Those studies which have investigated the impact of coping strategies on household livelihoods outcomes and impacts of unfavourable/insufficient livelihoods outcomes on coping strategies, have generally focused on food security as the livelihoods indicator.

Households were found to respond to food insecurity caused by shocks and stresses through adoption of various coping mechanisms, including reduction in quantity, composition and quality of foods consumed; collection of wild foods; reduction in daily meal frequency; borrowing from relatives, and inter-household food transfer, to name a few. Other studies have investigated the direct relationships between assets endowments and food security outcomes (e.g., Agbola et. al., 2004).

A great majority of those studies that have investigated the impacts of livelihood diversification on livelihood outcomes on the other hand have focused on per capita income or consumption as the main livelihoods indicator (e.g., Ersado, 2003; Jansen et. al., 2005). Several livelihoods diversification studies have found that the per capita incomes or consumption of households with more diverse livelihood strategies were less vulnerable to various shocks and stresses.

Therefore several researchers have argued that effective policies for poverty reduction would constitute those that encourage the diversification of livelihoods into various off farm and on farm activities. In fact, in the context of HPAI, Roland-Holst et al. (2008) have shown that policies based on livelihood diversification helps smallholders secure their future livelihood by sustaining their 'long-term risk management capacity' unlike a compensation scheme which creates negative coping strategies such as 'disease concealment' or sale of infected poultry.

Therefore the necessary institutions and incentives (such as subsidies, credit etc) should be provided to allow for rural households to diversify into various livelihoods activities. In order to investigate the impacts of HPAI on livelihoods strategies and outcomes, it is imperative to collect information on the various livelihood strategies rural households undertake and returns to each strategy, both before and after outbreaks; as well as the coping strategies they have adopted since the outbreaks and how these have changed their livelihoods outcomes.

H4: HPAI control policy affects household livelihood outcomes and household livelihood outcomes determine the efficacy of disease control policy.

Household livelihoods' rate of recovery from and/or resilience to HPAI outbreaks and threats depend on the efficacy and efficiency of the disease control policies. These policies include, but not limited to the following: timing and amount of compensation; extension services/ technical advice regarding poultry management in general and biosecurity measures in particular; subsidies to enable adoption of biosecurity measures, as well as availability of micro-credit to accumulate those assets necessary for more productive poultry production or participation in alternative livelihood strategies.

Furthermore, the success of disease control policy also depends on the livelihood outcomes (e.g., dependence on poultry for the overall household income, for food and nutrition security and gender equality), as well as household characteristics (e.g., risk perceptions and available assets which influence coping and risk management decisions) in addition to other policies (e.g., subsidies for crop production) and formal or informal institutions (e.g., banishment from the community or fines for non-reporting). Several livelihoods studies have investigated the impact of institutions and policies (such as food for work projects, currency devaluation, fluctuations in global commodity prices) on rural livelihoods. An overall review of the livelihoods studies reveal that while shocks and stresses such as natural disasters, droughts, and diseases are usually have negative impacts on the livelihood outcomes, impacts of policy and institutional changes on livelihoods outcomes could be either negative or positive.

There are also a few studies that have investigated theoretically the efficacy of HPAI control policy, especially the effectiveness of compensation schemes (e.g., Hennessy, 2005; Gramig et. al., 2005; Beach et. al., 2007b). These studies have shown that compensation schemes may create disincentives for reporting and/or for investment in biosecurity.

Other policies, including subsidizing private disease control measures and distribution of animal health information and education via agricultural extension or the media may provide incentives for more disease control, especially for smaller farms. Some authors (e.g. Otte et. al., 2006, Roland-Holst et. al., 2007) have also argued that compensation for HPAI outbreaks is an inappropriate intervention because it results in emergence of negative coping strategies (e.g. consumption and sale of infected birds, as explained above). Beach et al. (2007b) have further argued that making compensation depend on the level of biosecurity investment at least partially mitigates the disincentives generated by the compensation schemes, and other policies, such as subsidizing private disease control measures and distribution of animal health information and education via agricultural extension or the media may provide incentives for more disease control, especially for smaller farms.

In order to understand the relationship between livelihoods outcomes at the household levels and the HPAI control policies, information on the current disease control policies and prevailing formal and informal institutions should be collected. Qualitative and quantitative information on poultry producers' reactions to/perceptions of policies and institutions, as well as information on how these policies and institutions have evolved/are evolving, would also help explain the efficacy of these policies and institutions in HPAI control, as well as their impacts on livelihoods outcomes. The next section proposes inter-disciplinary and analytical methods, and as well as types of data, which could be used to test the four hypotheses introduced above.

6

Poultry Production and Health

Minnesota tops the nation in turkey production. The poultry industry brings in more than $800 million in income for producers, processors and other related industries. Our educational efforts translate into better bottom lines. We help the poultry industry improve management techniques, develop new products and find new markets.

This program provides information on: poultry production and processing; poultry diseases; nutrition and food safety; and poultry nutrition.

We serve a broad audience including the commercial poultry industry, niche producers, consumers and youth.

Nutritional and Feeding Strategies to Minimise Nutrient Losses in Livestock Manure

The livestock industry has undergone substantial changes in the past few decades. The poultry, swine, and dairy industries in particular have become increasingly concentrated. This has resulted in fewer, but larger, livestock operations throughout the country. As a result, there has been a growing concern about odor emissions from livestock operations and potential decreased water quality caused by nutrient runoff from livestock manure. One of the key components to lowering air and water pollution from manure is the animal diet.

All animals, whether ruminant (cattle and sheep) or non-ruminant (pigs and chickens), have five basic nutritional needs. Animals need energy which is provided by fats and carbohydrates in the diet, protein, water, vitamins, and minerals. Even in the best management situations, animals are not able to utilise 100% of the nutrients that are consumed.

Some nutrients will be excreted in the manure. Nutrients excreted in manure come from four primary sources: (1) feed wastage, (2) excess nutrients provided in the diet, (3) undigested nutrients in the diet, and (4) biological losses from dead cells in the body. From a manure management perspective two nutrients are of particular interest: *nitrogen* from the breakdown of protein, and the mineral *phosphorus*.

Nitrogen and Phosphorus

Nitrogen is essential in livestock diets. Proteins are made up of building blocks called amino acids, which in turn, contain the element nitrogen. Therefore, nitrogen is necessary to build protein. However, nitrogen excretion from animals is an environmental concern because it can volatilise into the air as ammonia causing offensive odours from livestock operations. Ammonia present in lakes and streams can be toxic to aquatic life. Nitrogen can also leach into ground water and cause pollution. Fifty percent of the total U.S. and 95% of rural residents obtain their drinking water from ground water sources.

Phosphorus is also necessary in animal diets. It is a key component of bone and teeth and is part of high energy compounds. Phosphorus is also an important part of cell membranes, RNA and DNA molecules. Phosphorus from animal manure binds to soil particles and can be carried by erosion to surface water such as lakes and rivers. Once in the surface water, phosphorus can cause excessive growth of algae. The algae use oxygen needed by fish, resulting in fish kills.

The average concentration of nitrogen and phosphorus in livestock and poultry manure. These are only averages. The actual concentration of nitrogen and phosphorus in the manure of a given livestock operation will vary with management, diet, manure storage facility, and age of animal housed in the facility.

Reduce Feed Waste

What can producers do to reduce the nitrogen and phosphorus that is excreted in livestock manure? One method of reducing nutrient losses in the manure is to *reduce feed wastage*. Feed wastage is any feed that is provided to, but not consumed by the animal. It can be feed that is left over in the trough or bunk or feed that is slopped over the edge of the feeder and ends up on the ground or in the pit. Van Heugten and Van Kempen (1999) found 2-12% feed wastage in the U.S. swine industry. They estimated that if just 5% of feed is wasted, a net income loss of $1.77/pig depending on the market can

result and an additional 327 grams of nitrogen and 82 grams of phosphorus is excreted per pig. Producers can reduce feed wastage in their livestock operation by regularly adjusting feeders, using proper feeder design for age and type of animal, and regular maintenance of the feeders or bunks. The use of total mixed rations (TMR) for cattle and pelleted feed for swine and poultry can also reduce feed wastage.

Increase Accuracy

A second method of reducing nutrient losses in manure is to *more accurately meet the nutrient requirements of the animal.* Animals are only able to utilise a certain amount of any given nutrient for maintenance and productive purposes. Nutrients supplied above that level are excreted by the animal.

Therefore, while it is important that all animals are supplied sufficient amounts of each needed nutrient, it is also critical that producers do not grossly over-supply nitrogen and phosphorus in the diet. Male and female animals do not have the same nutrient requirements. Nutrient requirements also change as the animal progresses from one phase of production to another. For example, the nutrient needs of starter calves are not the same as the nutrient needs of feeder cattle. Therefore, animals should be grouped by gender and phase of production and fed diets specific to their nutrient needs. Feeding animals in groups according to their gender is known as split-sex feeding. Feeding according to phase of production is known as phase-feeding.

In the past, many producers added nutrients at levels above the recommended level necessary for maintenance and production as a "safety margin." Spears (1996) found that the typical diet for North Carolina finishers contained 0.40 to 0.70% total dietary phosphorus. The recommended phosphorus level for finisher pigs was only 0.40%. On average, producers were supplying their pigs with 1.55 times the phosphorus necessary for growth and maintenance.

A similar study conducted with Wisconsin dairy cattle revealed that producers were providing 0.45 to 0.50% total phosphorus in the diet when the recommended level of total phosphorus for a dairy cow was only 0.38%. Excess phosphorus provided to livestock can not be utilised by the animals and is excreted in the manure. Simply lowering the dietary phosphorus level 0.1% will decrease phosphorus excretion about 8.3% in pigs (Kornegay and Verstegen, 2001). Due to research studies such as these, many producers now lower safety margins to reduce nutrient excretion and save money.

Another way to more accurately meet the nutrient requirements of the animal is to use synthetic amino acids in poultry and swine diets. Synthetic amino acids are manufactured sources of highly concentrated specific amino acids. Traditionally, high protein ingredients such as soybean meal or meat and bone meal were used to supply the animal with necessary protein and amino acids. But these feed ingredients provide a wide variety of amino acids, not always in the specific quantities needed by the animal. When synthetic amino acids are supplemented into the diet, less high protein feed ingredients are needed to meet to the amino acid requirement of the animal, and the amino acids are added at levels more closely reflecting the nutrient needs of the animal. This reduces the total protein content of the diet and decreases nitrogen excretion in manure. For every 1% reduction in crude protein achieved by using synthetic amino acids, total nitrogen losses are reduced by 8% (Kerr and Easter, 1995). Reducing nitrogen excretion also decreases ammonia production and may lower odor emissions from the livestock facility.

Improve Digestibility

A third method of reducing nutrient losses in manure is to *improve the digestibility of the diet.* When a highly digestible diet is fed, a higher percentage of the nutrients are used by the animal for maintenance and growth and fewer nutrients are excreted in manure. Some feed ingredients are more digestible than others. For example only 14% of the phosphorus in corn is actually in a form that is available to a pig, while 90% of the phosphorus in corn distiller's dried grains with solubles is available to a pig. By selecting feed ingredients that are highly digestible, producers can reduce the nutrients excreted in manure. However, producers need to balance digestibility of the diet with economics when deciding which feed ingredients best fit into their operation.

An increasingly common way to improve phosphorus digestibility in swine and poultry diets is to add a commercially available enzyme called phytase to the diet. Most phosphorus in cereal grains fed to livestock is bound to a compound called phytate. As a result, the phosphorus is not available to the animal for growth and maintenance purposes and is excreted in the manure. Phytase is an enzyme that breaks the phytate compound and increases phosphorus availability. As a result, more of the phosphorus from cereal grains is available to the animal and less is excreted in the manure. Phosphorus excretion can be reduced 33% when phytase is added to a low phosphorus diet

(Kornegay and Verstegen, 2001). It is not necessary to add phytase to cattle diets because the microbes in their stomach naturally produce phytase.

There is a variety of corn called low-phytate corn that has the same effect as adding phytase to swine and poultry diets. This corn variety has the same total phosphorus as conventional varieties, but low-phytate corn naturally has less total phosphorus bound to the phytate compound. Consequently, more phosphorus is available to the pig or chicken. Currently, it is more economical for producers to feed conventional corn with supplemental commercial phytase than to use low-phytate corn in the diet. This may change with continued improvement in plant genetics.

The implementation of nutrition and feed strategies discussed above will significantly reduce nutrient excretion, but they are not the total solution to the amount of manure produced by livestock, since 100% utilisation of a diet is not possible. As we look ahead to the future, there may be other dietary strategies that become available to assist with lowering nitrogen and phosphorus excretion in livestock manure. However, economics will always be a driving factor behind producer acceptance of these new technologies since the producer must maintain a profit to stay in business. Therefore, continued research is needed is this area to ensure that producers have economical technologies available to help lower nutrient excretion from their livestock facility.

The Effects of Plant Extracts on Bacteria by Molecular Profiling

NEXT Enhance® 150 is a standardised combination of thymol and cinnamaldehyde, an active ingredient of cinnamon extract. This combination acts to optimise gut flora equilibrium in the broiler and breeder chicken, leading to better feed efficiency and growth for the animal.

Continuing to understand the mechanisms behind the consistent benefits in animal trials, Carotenoid Technologies SA (CaroTech®) has worked exclusively with the Zootechnical Department at the University of Ljubljana in Slovenia., one of the few sites in Europe which has the expertise in the state-of-the-art methodology of PCR-DGGE fingerprinting. Previous work by the two groups has elucidated the effects of the combination of thymol and carvacrol found in Origanum spp. have been shown by molecular profiling to increase Lactobacillus populations and reduce E. coli.

New scientific results have been published in Folia Microbiologica 53 (3), 204-208 (2008). Entitled *"Effect of Sodium Monensin and Cinnamaldehyde on the Growth and Phenotypic Characteristics of Prevotella bryantii and Prevotella ruminicola"* (D. Ferme, M. Malnersic, L. Lipoglavsek, C. Kamel, G. Avgustin).

Current studies are underway with a European partner to determine the effects on Clostridium spp. and other pathogens. These novel findings not only pave the way for a better understanding of the active ingredients, but can be used as supportive data in the product registration as a zootechnical additive under European Directive 1831/2003.

Nutrition Issues Challenge Broiler Performance

Feed quality has become a formidable obstacle to achieving broiler performance goals, according to poultry nutritionists and consultants.

"Our greatest challenge is ingredient quality," says Dr. Tom Frost, director of nutrition and research, Wayne Farms, Oakwood, Georgia.

"We continue to receive increased pressure to use more and more byproduct ingredients, yet the quality of those ingredients continues to go down."

Poultry meal is a prime example. With more of it being diverted to the pet food industry, the meal that's left is far from the quality it was before, and the same holds true for meat- and bone-meal blends and fat. "The rendering situation — it's a challenge," he says.

Another example is distillers' dried grains (DDG). It hit the industry as a prime ingredient, but the ingredients and fat content need scrutiny. "This, too, needs to be stabilised," adds Frost, who shared his views with other poultry nutritionists who came together for a roundtable discussion, "Feeding Broilers for Optimum Health, Performance and Profitability." The event was sponsored by Intervet/Schering-Plough Animal Health.

DDG Market 'Immature'

Bill Johnson, senior nutritionist with Tyson Foods, Magee, Mississippi, says that one disadvantage to the new DDG market is that it's very immature and doesn't yet have well-established trading rules. Much DDG is traded through brokers and resellers, not by direct sale from the producers.

Strong agreement came from Dr. Paul Twining, independent poultry consultant, Princess Anne, Maryland. A primary obstacle is determining the nutrient value for ingredients coming into the feed mill, and as for DDG, "you've got to have information going clear back to the production plant. There can be a tremendous difference between DDG from plant A versus plant B.

"Personally, I won't buy DDG on the open market, through a broker. I want to know exactly the plant of origin. It's the same with bakery meal," he says.

Dr. Shivaram Rao, director of nutrition and feed mills for Foster Farms, Livingston, California, thinks it takes a lot of homework when alternative ingredients such as DDG are brought in.

"We've got to make sure that we know the consistency and digestibility of the product so that we are not altering the amino acid balance or the calorie-to-protein ratio in the finished feed," he says.

Current lab techniques are an added frustration, Rao adds, because results come back late. By the time moisture or protein changes are detected, the birds have already eaten that feed, he says.

Dr. Robert Teeter, professor of animal science, Oklahoma State University, Stillwater, says that the calorie and protein package — as well as coccidiosis control — are "extremely important" to broiler performance, but so are other aspects of the managerial environment that have a calorific cost. Lighting programs today, for instance, are saving energy on the order of about 100 kcals per kilogram. Pellet quality is another managerial factor that will increase the effective caloric value of rations.

Additional Complications

Several other factors complicate the job of the poultry nutritionist, roundtable participants said. One is extreme variability in corn quality, depending on where it was grown. "We've seen considerable differences in mycotoxin levels in the corn," Twining says. "And there may be some fairly extreme differences in the caloric content of corn that's not showing up in our standard approximate analysis." Weather as well as housing conditions can pose additional problems that hinder broiler performance goals. "We have to make sure that the growers are on our program," he adds Dr. Mike Blair, nutritionist, Pilgrim's Pride, Woodstock, Georgia, pointed to feed mills as a limiting factor in poultry nutrition. Bin space, especially with older mills, is a challenge and another is getting people to pay attention to how they're feeding the birds. Good communication with live service techs is imperative, he says. Looming restrictions on in-feed antibiotics is another significant challenge facing poultry nutritionists.

Although the antibiotic-free market isn't here now, Rao says it's well on the way. "In a year or two, it's going to come," he predicts. "I'm concerned that we be prepared for that — that we be a source of antibiotic-free poultry." Rao adds: "I'm kind of excited about vaccines as an alternative to some in-feed antibiotics. They're something we need to keep improving on and maybe they will give us some answers."

Lessons from Antibiotic-free Flocks

The participants acknowledged there have already been lessons learned from antibiotic-free birds that they've applied to conventional flocks.

According to Twining, one lesson is that better husbandry results in fewer problems. "We've seen lower bird density in the antibiotic-free programs, and we're using some fairly high levels of probiotics in place of an antibiotic," he says. "All of these things are expensive in a large-scale production system, so it comes back to a speciality-type program where we have to have a premium to make it economical."

At Foster Farms, some management strategies from its antibiotic-free/organic program have been applied to the conventional flocks. "Mainly it's the fat quality," Rao insists. "In our antibiotic-free/organic program, we use high-quality soybean oil and the performance is better than in our conventional program. So now we have our own fat-blending facility. We have dramatically improved our fat quality."

Foster Farms has also learned that it's important to provide more space. Based on results from the antibiotic-free/organic program, "we were able to convince management to give us more space for the conventional birds," Rao adds. Occasionally, additional lines not belonging to the pattern for the zeolite appear in a pattern along with the X-ray lines characteristic of that zeolite. This is an indication that one or more additional crystalline materials are mixed with the zeolite in the sample being tested. Small changes in line positions may also occur under these conditions. Such changes in no way hinder the identification of the X-ray patterns as belonging to the zeolite.

The particular X-ray technique and/or apparatus employed, the humidity, the temperature, the orientation of the powder crystals and other variables, all of which are well known and understood to those skilled in the art of X-ray crystallography or diffraction can cause some variations in the intensities and positions of the lines. These changes, even in those few instances where they become large, pose no problem to the skilled X-ray crystallographer in establishing identities. Thus, the X-ray data given herein to identify the lattice for a zeolite, are not to exclude those materials which, due to some variable mentioned or otherwise known to those skilled in the art, fail to show all of the lines, or show a few extra ones that are permissible in the cubic system of that zeolite, or show a slight shift in position of the lines, so as to give a slightly larger or smaller lattice parameter.

A simpler test described in "American Mineralogist," Vol. 28, page 545, 1943, permits a quick check of the silicon to aluminium ratio of the zeolite. According to the description of the test, zeolite minerals with a three-dimensional network that contains aluminium and silicon atoms in an atomic ratio of Al/Si=2/3=0.67, or greater, produce a gel when treated with hydrochloric acid. Zeolites having smaller aluminium to silicon ratios disintegrate in the presence of hydrochloric acid and precipitate silica. These tests were developed with natural zeolites and may vary slightly when applied to synthetic types.

U.S. Pat. No. 2,882,243 describes a process for making zeolite A comprising preparing a sodium-aluminium-silicate water mixture having an SiO.sub.2 :Al.sub.2 O.sub.3 mole ratio of from 0.5:1 to 1.5:1, and Na.sub.2 O/SiO.sub.2 mole ratio of from 0.8:1 to 3:1, and an H.sub.2 O/Na.sub.2 O mole ratio of from 35:1 to 200:1, maintaining the mixture at a temperature of from 20.degree. C. to 175.degree. C. until zeolite A is formed, and separating the zeolite A from the mother liquor.

As noted above, this invention involves, inter alia, a method for reducing degradation of internal egg quality. In one of its forms, this method involves a process for obtaining an egg crop having a reduced rate of internal egg quality degradation which comprises (i) preparing a poultry diet which contains from about 0.25 to about 3.0 weight percent zeolite A, (ii) feeding such diet to laying poultry, and (iii) recovering an egg crop therefrom in which the rate in decrease in Haugh unit value of the eggs is reduced as compared to the rate of such decrease under the same conditions with the same diet absent the zeolite A.

In another of its forms, the present invention relates to a method of improving the egg shell characteristics, e.g. egg shell strength of eggs from laying poultry. A convenient means of measuring egg shell strength is by measuring the specific gravity of the egg.

The present invention is in the general field of poultry farming and relates particularly to the feeding of laying fowl or layers. In a particular aspect, it relates to a method of feeding female birds with a feed formulation which enhances, for example, interior egg quality, as well as egg production, and/or egg shell quality. This invention also relates to the eggs with improved egg shells obtained from birds fed a diet of this invention. Poultry diets of this invention are typically diets which include a zeolite such as zeolite A in an amount such that degradation of the interior quality of eggs is retarded. This invention also relates to improved poultry farming. Thus, it relates to methods which provide improved egg production. These methods comprise obtaining an increase in yield of marketable eggs from laying hens. The increase is provided by feeding hens a diet of this invention.

Description of Related Art

The literature associated with poultry nutrition, and the effect of both nutrition and egg processing on egg quality is so vast as to defy brief summation. A few of the developments in this field are highlighted below.

In previously issued patents, we disclosed that zeolite A enhances egg shell quality, U.S. Pat. No. 4,556,564; improves feed utilisation efficiency, U.S. Pat. No. 4,610,882; and improves liveability, U.S. Pat. No. 4,610,883. The disclosures in those patents are incorporated by reference herein as if fully set forth. Other related art is mentioned in the patents.

An article by C. Y. Chung et al from Nongsa Sihom Young Pogo 1978, 20 (Livestock) pp. 77-83 discusses the effects of cation exchange capacity and particle size of zeolites on the growth, feed efficiency and feed materials utilisability of broilers or broiling size chickens. Supplementing the feed of the broilers with naturally occurring zeolites, such as clinoptilolite, some increase in body weight gain was determined. Chung et al also reported that earlier results at the Livestock Experiment Station (1974, 1975, 1976—Suweon, Korea) showed that no significant difference was observed when 1.5, 3, and 4.5 percent zeolite was added to chicken layer diets.

U.S. Pat. No. 3,836,676 issued to Chukei Komakine in 1974 discloses the use of zeolites and ferrous sulfate crystals in an odourless chicken feed comprising such crystals and chicken droppings. The results were said to be no less than those in the case where chickens were raised with ordinary feed.

Experiments have been in progress in Japan since 1965 on the use of natural zeolite minerals as dietary supplements for poultry, swine and cattle. Significant increases in body weight per unit of feed consumed and in the general health of the animals was reported (Minato, Hideo, Koatsugasu 5:536, 1968). Reductions in malodor were also noted.

Using clinoptilolite and mordenite from northern Japan, Onagi, T. (Rept. Yamagata Stock Raising Inst. 7, 1966) found that Leghorn chickens required less food and water and gained as much weight in a two-week trial as birds receiving a control diet. No adverse effects on health or mortality were noted. The foregoing Japanese experiments were reported by F. A. Mumpton and P. H. Fishman in the Journal of Animal Science, Vol. 45, No. 5 (1977) pp. 1188-1203.

Canadian Pat. No. 939,186 issued to White et al in 1974 discloses the use of zeolites having exchangeable cations as a feed component in the feeding of urea or biuret non-protein (NPR) compounds to ruminants, such as cattle, sheep and goats. Natural and synthetic as well as crystalline and non-crystalline zeolites are disclosed. Zeolites tested included natural zeolites, chabazite and clinoptilolite and synthetic zeolites X, Y, F, J, M, Z, and A. Zeolite F was by far the most outstanding and zeolite A was substantially ineffective.

In a study at the University of Georgia, both broilers and layers were fed small amounts (about 2%) of clinoptilolite, a naturally occurring zeolite from Tilden, Tex. The egg shells from the hens receiving zeolite were slightly more flexible as measured by deformation, slightly less strong as measured by Instron breaking strength, and had a slightly lower specific gravity. The differences in egg shell quality were very small. This type of zeolite was ineffective in producing a stronger egg shell. An article written by Larry Vest and John Shutze entitled "The Influence of Feeding Zeolites to Poultry Under Field Conditions" summarising the studies was presented at Zeo-Agriculture '82.

A study by H. S. Nakone of feeding White Leghorn layers clinoptilolite, reported in 1981 Poultry Science 60:944-949, disclosed no significant differences in egg shell strength between hens receiving the zeolite and hens not receiving the zeolite.

Background of the Invention

As is well known, the interior quality of eggs is diminished with time. The loss in quality attributes of the albumen and yoke is a function of temperature and movement of carbon dioxide through the shell. Low temperatures decrease the rate of loss in Haugh units—a standard measure of interior egg quality—and thus it is recommended that eggs be stored at temperatures close to the freezing point, a procedure which, as a practical matter, is not always feasible. To reduce rate of carbon dioxide (and moisture) loss various shell treatments have been utilised, such as spraying oil on the eggs.

A way has been discovered for reducing, if not eliminating, the need for refrigeration and spray oiling on eggs as a means of preserving interior egg quality.

To ameliorate the problem of interior egg degradation with time, the following procedures have been recommended heretofore:

1. Gather eggs three to four times per day.
2. Clean the eggs promptly after gathering and cool for 12-24 hours at 13.degree. C. or preferably 10.degree. C. before packing in cases or cartons.
3. Keep the eggs at between 60 to 85%, preferably 70 to 80%, relative humidity.
4. Resort to careful handling.
5. Use proper packing using precooled containers only.
6. Resort to frequent marketing of not less than twice a week.
7. Use speedy, refrigerated transportation and make frequent deliveries to sales outlets, preferably at least five times per week.
8. Use adequately refrigerated holding spaces at sales outlets.
9. Keep the eggs in home refrigerators at 7.degree. C. to 13.degree. C., and preferably use all of the eggs within one week.

Summary of the Invention

The demand for poultry eggs, especially chicken eggs expanded considerably over the last decade. The poultry industry has grown from a home industry to a large scale manufacturing industry in which tens of thousands of eggs are produced daily at single farms or egg laying installations. Some eggs are produced for eating and some eggs are produced for hatching. One problem with such large scale egg producing is premature reduction in interior egg quality as a function of time. That is, unless eggs are handled and/or treated in accordance with the above 9-point program, their quality as regards internal qualities of the albumen and yolk may deteriorate faster than would be desired. Moreover, a way of simplifying egg production, distribution, marketing, etc. with concomittant reduction in refrigeration requirements would be a welcome contribution to the art. Another problem associated with large scale egg production is breakage. Even a slight crack in an egg makes it unsuitable for hatching and most other marketing purposes. It is estimated that some six percent of all eggs produced are lost for marketing because of cracking or breakage. Shell strength is very important to inhibit breakage. The stronger the egg shell, the less likely the egg will be cracked or broken. Machinery and techniques necessary for carefully handling the eggs to avoid breakage are expensive and time consuming.

Another substantial loss of egg production, estimated to be about a seven percent loss, is the production of shell-less eggs. Any reduction in shell-less eggs can be an important factor in large scale egg production.

We have discovered that internal egg quality can be maintained at a higher level by feeding laying poultry a diet containing zeolite A. In other words, this invention provides, in one of its embodiments, a process for reducing the rate of internal egg quality degradation which comprises feeding a laying poultry hen a diet which contains zeolite A such that the rate in decrease in Haugh unit value of the eggs from said hen is reduced as compared to the rate of such decrease under the same conditions with the same diet absent the zeolite A. In this process, the amount of zeolite A is usually within the range of about 0.25 to about 3.5 weight percent, preferably between about 0.75 to about 1.5 weight percent.

As stated above, our previously issued U.S. Pat. No. 4,556,564 discloses that zeolite A in poultry feed improves egg shell quality.

In Nutrient Requirements of Poultry, Eighth Revised Edition 1984, National Academy Press, Washington, D.C. (1984) it is state, that leghorn-type chickens, both layers and breeders, have a dietary requirement of 0.15 weight % chlorine per day. Assuming an average daily intake per hen of 110 g of feed per day, this amounts to 165 mg of chlorine, per hen per day. This requirement is usually furnished as chloride ion, typically in salt, i.e. NaCl.

In the practice of this invention, the chloride level of the poultry diet may be reduced in order to improve egg yolk production and/or egg shell quality. At the same time, the utilisation of such a diet achieves improved retention of internal egg quality as manifested by egg crops in which the rate in decrease in Haugh unit value of the eggs is reduced as compared to the rate of such decrease under the same conditions with the same diet absent the zeolite A. Accordingly, we have discovered that egg production, egg quality, notably internal egg quality, and/or egg shell quality of poultry eggs is enhanced if, besides adding a zeolite (notably zeolite A) to the diet, the chloride level in poultry diets is decreased below the previously recommended requirement. The improvement can result in the production of more collectable eggs, which with broiler breeder hens means more eggs that are settable in incubators.

The enhancement of results provided by addition of zeolite A is unexpected. Furthermore, reduction of chloride levels below the previously required amount is unexpected.

Description of Preferred Embodiments

This invention comprises a method for improving the internal quality of eggs in terms of preservation of Haugh units. This method involves, in another embodiment, a process for obtaining an egg crop having a reduced rate of internal egg quality degradation which comprises (i) feeding laying poultry a diet which contains from about 0.25 to about 3.0 weight percent zeolite A, and (ii) recovering an egg crop therefrom in which the rate in decrease in Haugh unit value of the eggs is reduced as compared to the rate of such decrease under the same conditions with the same diet absent the zeolite A. In addition, this invention comprises a method for improving eggs, egg production and/or egg shell quality. This method comprises feeding birds a diet which (a) contains reduced amounts of chlorine, e.g. less chloride than what was previously recognised as required, and which (b) also contains a zeolite such as zeolite A. Thus, in an important aspect, this invention in one of its forms can be considered a method of enhancing the quality of eggs, both in interior quality and in egg shell quality by use of a zeolite.

In still another preferred embodiment, this invention provides an egg produced by laying poultry fed a diet which contains zeolite A, the egg being characterised by exhibiting a rate of decrease in Haugh unit value that is lower than the rate of such decrease under the same conditions with the same diet absent the zeolite A. Preferably, these eggs are produced by a chicken hen, although the invention is deemed applicable to comparable improvements in duck eggs, quail eggs, turkey eggs, and the like.

Zeolites are crystalline, hydrated aluminosilicates of alkali and alkaline earth cations, having infinite, three-dimensional structures.

Zeolites consist basically of a three-dimensional framework of SiO.sub.4 and AlO.sub.4 tetrahedra. The tetrahedra are cross-linked by the sharing of oxygen atoms so that the ratio of oxygen atoms to the total of the aluminium and silicon atoms is equal to two or O/(Al+Si)=2. The electrovalence of each tetrahedra containing aluminium is balanced by the inclusion in the crystal of a cation, for example, a sodium ion. This balance may be expressed by the formula Al/Na=1. The spaces between the tetrahedra are occupied by water molecules prior to dehydration.

Zeolite A may be distinguished from other zeolites and silicates on the basis of their composition and X-ray powder diffraction patterns and certain physical characteristics. The X-ray patterns for these zeolites are described below. The composition and density are among the characteristics which have been found to be important in identifying these zeolites. The basic formula for all crystalline sodium zeolites may be represented as follows:

Na.sub.2 O.Al.sub.2 O.sub.3.xSiO.sub.2.yH.sub.2 O

In general, a particular crystalline zeolite will have values for "x" and "y" that fall in a definite range. The value "x" for a particular zeolite will vary somewhat since the aluminium atoms and the silicon atoms occupy essentially equivalent positions in the lattice. Minor variations in the relative number of these atoms do not significantly alter the crystal structure or physical properties of the zeolite. For zeolite A, the "x" value normally falls within the range 1.85.+-.0.5.

The value for "y" is not necessarily an invariant for all samples of zeolites. This is true because various exchangeable ions are of different size, and, since there is no major change in the crystal lattice dimensions upon ion exchange, the space available in the pores of the zeolite to accomodation.

This is simply done by immersing the egg in solutions of salt water of varying strengths. It is well known in the art that specific gravity correlates with egg shell strength. As specific gravity of the egg is raised, the strength of the egg shell is increased.

This latter method of this invention can also improve egg shell thickness, and/or decrease the number of eggs produced without shells.

The improvements of this invention are obtained by adding a zeolite to the diet of laying poultry. The zeolite can be added to regular or standard poultry feed or administered directly to the birds by some other means. The percent of zeolite (such as zeolite A) in the diet is typically from about 0.25 to about 3 or 4 weight percent, as discussed below. For this invention, chloride intake of the birds is reduced, even below the levels previously recognised as required. For leghorn-type poultry the chlorine level in the diets used in this invention are below 0.15 weight percent; more preferably, between about 0.04 and 0.10 weight percent. The benefits of this invention can be obtained with ETHACAL.RTM. feed component, a commercially available form of sodium zeolite A.

As indicated above, Zeolite A is added to such feed formulation in small amounts by weight percent of up to about four weight percent. Greater amounts may be used, but may deprive the poultry of the desired amount of nutrients. Greater amounts are also likely to be cost ineffective. A preferred amount of zeolite A is from about one-half to about two percent by weight of the total feed formulation. A most preferred amount of zeolite A is from about 0.75 to about 1.50 weight percent of the total feed formulation.

For this invention the term "poultry" includes all domestic fowl, namely chickens, turkeys, ducks, geese, and the Corn is the principal diet for most laying poultry. A feed formulation comprising by weight percent the following is desirable

Summary of Results

Examples 1 and 2 demonstrate quite clearly that use of zeolite A in poultry feed improves the resistance of the eggs to internal quality degradation over time. Besides enhancing egg quality during storage and transportation, these results are of importance in that the longer hatching eggs are stored prior to incubation, the lower will be the percentage of hatch. Thus, the invention provides eggs with superior characteristics both as regards "freshness" for human consumption, and enhanced hatchability. Indeed, some published studies indicate that chicks hatched from eggs stored for over 8-10 days perform more poorly when fed to market weights. Thus, improvements in these characteristics should likewise be achieved by utilisation of this invention.

In Example 4 hens fed the zeolite A, low chloride diet laid eggs with the highest specific gravity, shell thickness and percent shell, and they produced the lowest percentage of shell-less or membrane eggs, Hens fed the zeolite A diets containing high chloride, equivalent to the chloride level of the control diet (Diet 1), also laid eggs with the same improvements, but of lesser magnitude than the low chloride fed hens. These improvements result in the production of more collectable eggs, which with broiler breeder hens means more eggs that are settable in incubators. That inclusion of zeolite A in the diets have no adverse effect on overall egg production or feed consumption. That inclusion of zeolite in the diet results in increased weight gain suggesting that improved feed utilisation is occurring.

The above results suggest the use of zeolite A levels in the diet of from about 0.25 to about 3.0 weight percent, or up to about 4.0 weight percent. The above results also suggest that for chickens, both layers and broiler breeders, the chlorine content (% chloride) in the diet can be less than about 0.15 weight percent. Preferably the chloride content is from about 0.04 to about 0.10 weight percent.

The results set forth in the Examples above demonstrate the advantages of this invention in improving interior egg quality, egg shell quality, shell thickness, percent shell, or in reducing the number of shell-less or membrane eggs. Any such benefit other than improved interior egg quality, for the purposes of this invention is referred to as an improved egg shell characteristic. The above results demonstrate an improved process for enhancing either or both of (i) interior egg quality and (ii) an egg shell characteristic which comprises regularly feeding to laying poultry a diet which comprises a zeolite such as zeolite A. To achieve the benefits of (ii) the diet also contains a reduced chlorine content. As shown, the diets with reduced chlorine content can contain an amount of chlorine (as chloride) which is below that which was previously recognised as required. Such diets are referred to herein as chloride deficient diets. These diets can comprise the chloride content supplied by the corn and soybean meal or similar portions of the diet without any added chloride source (such as salt, i.e. NaCl), or with an amount of salt or other chloride below that which is commonly added to the diet, as illustrated above.

Generally speaking, one uses enough zeolite such as zeolite A to provide the benefit or benefits conferred by such additive. Preferred ranges are set forth above. Besides processes comprising feeding birds, this invention also provides chloride deficient diets or feed formulations which have zeolite and chloride contents as discussed above.

Preferred aspects of this invention comprise use of zeolite A diets (including chloride deficient diets of zeolite A) for poultry, although diets of this invention can be used in other aspects of animal husbandry such as in the care and feeding of endangered, wild species. Of the poultry species, chickens, including layers and broiler breeders are highly preferred. Use of the diets of this invention provides improved methods of poultry farming which can result in more eggs that are usable, and which have desirable characteristics.

Having described this invention and its benefits in detail above, it will be apparent that those skilled in the art can make modifications and changes of the invention as above-described without departing from the spirit and scope of the claims which follow.

Contributions from/by Livestock-keepers

Subsidies can become a mechanism for securing farmers' involvement in trials and treatments that they do not really consider to be worthwhile: for this reason they should be avoided or minimised. On the other hand, there are potential problems in avoiding them altogether from the outset. First, this may make it difficult for resource-poor livestock-keepers to participate in trials, particularly where the treatment has to be purchased. Second, many rural people have a dependency mentality, having become accustomed to receiving government handouts, and hence are reluctant to pay for things themselves when they are working with development agencies. This is particularly the case in India.

Unless the livestock-keepers have a strong sense of ownership of the trial, and believe that it is very important and has a reasonable chance of success and profitability, they are unlikely to pay for the whole treatment themselves, particularly if it is one that they are not very familiar with. This is illustrated by the experience of a de-worming trial in Kenya, The treatments used in the trials conducted by the BAIF/NRI project were subsidised to varying degrees.

The basis for this was that the technologies were new to the goat-keepers, and that they were therefore taking a risk (financial and potentially to the health of their goats) in applying them. Urea Molasses Granules was the newest of all the treatments, so a 100% grant was given for this. In this kind of situation the researchers should develop a clear understanding with the livestock-keepers that the size of the subsidy will be reduced, year by year, as the goat-keepers become familiar with the technologies and see the benefits they have on their animals (assuming, that is, that they are effective!).

Incentives for Participants in Control Groups

For most kinds of trials it is desirable to have a group of animals that do not receive any treatment, so that comparisons can be made between the two groups to see what difference the treatment made. The non-treatment group is known as the 'control' group. If the owners of the control group animals also own the treatment group animals, motivation is not likely to be a problem. However, if they are different to those of the treatment group motivating the control group owners to participate actively in the control group can be a challenge. They may be reluctant to participate, because they do not see themselves or their animals gaining anything from their involvement in the trial.

If this situation arises, it should be pointed out to them that if the technology on trial proves to be effective they too will benefit after the trial has been completed. If that is not enough of an incentive, however, they can be offered some kind of material incentive. For example, in the BAIF/NRI project control group goat-keepers participating in a trial focusing on milk yield were given metal drink containers; and in other trials a breeding buck was made available to both the treatment and control groups.

Apparently, ICRAF do not carry out on-farm feeding trials comparing treatments and a control, only different treatments, partly because of the issues raised (Morton, 2001). We would argue strongly in favour of having a control group, and finding ways of motivating them, such as those just mentioned. They are just as important to a trial as the treatment group members, so researchers should put just as much effort into developing their sense of ownership and involve them fully in group meetings, etc.. If there is a need to hire the services of a local monitor for monitoring the trial, preference can be given to someone from the control group.

Human Resources

Researchers' attitudes are very important. They should have respect for the views and knowledge of the people with whom they intend to work. They should be prepared, for example, to traverse difficult terrain and climates, leave home early and return late, spend time in the project area, and hold interviews at times that are convenient for local people. This is particularly important when working with women.

FPR/PTD requires staff skills that are often scarce. Many researchers in national agricultural research organisations may lack experience and aptitude in working with farmers in a participatory way. While many NGO staff may have little, if any, experience The precise nature of the disciplines will depend to some extent on the constraints and opportunities that livestock-keepers identify as being most important. For example, if heavy worm burdens appears to be a serious problem, a parasitologist may be required to identify which helminths are involved and to advise on an effective treatment.

Alternatively, if the focus were on improved forage production, then a forage agronomist and livestock nutritionist might be needed. However, the basic minimum should be one livestock scientist and one social scientist or agricultural economist.

Ideally, at least one member of the research team should have had some previous experience of participatory research. In addition, it is generally desirable, and sometimes essential, to have at least one woman in the team, to facilitate interaction with female livestock-keepers. There is often a need for training of research staff in participatory research methods. To some extent it may be possible to provide 'on-the-job' training, but if funds permit short (2-day to 2-week) formal courses would be ideal..)

Financial Resources

The financial resources required for collaborative research are often underestimated. Monitoring and analysis may be more time-consuming than they are in more conventional research modes. With on-farm experimentation there is generally more variability in the experimental data than is the case with the contract mode, and consequently data may need to be obtained from a larger number of fields, farmers, etc., if the effect of the technology being tested is to be detected. It may also be necessary to carry out more detailed monitoring of farmers' management activities to explain some of the variability in results between herds. It will also take time, and hence money, to build-up an effective working relationship.

Research projects normally have a fixed duration, which is usually no more than 3 years. Research scientists need to consider whether FPR can deliver/meet objectives in the time earmarked for the project, while bearing in mind the need to (a) allow for delays and complications, and (b) develop a rapport and partnership with participants. For example, if they have not worked closely with local communities already a long lead-in time may be required to develop the necessary rapport to work effectively with them.

Any type of livestock research, whether participatory or not, should be based on a sound understanding of what livestock-keepers see as their priority needs - their main constraints or opportunities. In the case of PTD this is an essential foundation for the whole process, since people are unlikely to invest time and effort on research that they consider to be unimportant.

Obtaining an accurate understanding of needs and priorities can be difficult and time-consuming, and may require at least two phases of discussions with farmers. Directly asking people their most pressing problems may merely generate well-known 'shopping lists'. Problems are likely to be described in terms of a lack of an input (which the farmer hopes the project will provide): for example, where there is a high mortality rate, livestock-keepers may characterise the problem as a "lack of veterinary medicines". It is important to identify the underlying cause of the problem, rather than just the symptoms. Getting farmers to rank problems or priorities, starting with the most serious or important, provides more information than simply making a list; it also reduces the risk of researchers distorting farmers' views to fit in with their own personal interests or priorities. A companion guide to this one, entitled Participatory Situation Analysis with Livestock-Keepers: A Guide (Conroy, 2001), provides valuable guidance on conducting needs assessments with livestock-keepers, including the use of problem-trees to identify underlying causes. Researchers should seek to increase their understanding of what is required and possible as the research progresses.

Ideally, the researchers assist livestock-keepers in tackling their most pressing production problems. However, not all of these problems can be easily solved by improved technologies (although there may still be scope for influencing policies or institutions) and their research institute may not have the expertise (or resources and willingness) to address certain problems. Nevertheless, so long as the issue is a reasonably high priority for the local people, and researchers are also convinced, there is likely to be potential for fruitful collaboration.

The broader the scope of the project, and the greater the variability of the systems and situations of the target group(s), the more time and effort will be required for the needs assessment exercise.

Scheduling

The timetable for the research needs careful consideration. There are often conflicting demands regarding the pace at which research progresses. The short duration of many research projects (often no more than 3 years) may encourage researchers to establish on-farm trials as quickly as possible, so that they can gather data for a minimum number of seasons or years.

On the other hand, it is desirable to avoid rushing because this can undermine the participatory approach as farmers/users are not given an equal say in the design of the research, and they may not develop a sense of ownership. It may be desirable to focus on interacting with fewer farmers and villages during the first season or year so that staff become familiar with a participatory approach. This is a time when effective relationships are developed with participants, research opportunities are identified, and the technologies to address them are agreed. It is important that the research team members are not over-ambitious and do not spread themselves too thinly. A narrowly focused approach will be preferable where:

- resources are limited;
- staff are relatively new to PTD; or
- a lot of time needs to be invested in building up rapport with the communities.

Where the approved duration of the research is longer (say up to 5 years), it may be best to have an initial phase of observing farmers' own informal trials or current practices, in relation to selected issues. This helps the researchers understand problems and potential solutions from the farmers' viewpoint, gain deeper insight into the differing problems of individual farmers or sub-groups, and prepares them for participatory research if they are new to it.

Observations of this kind can be conducted concurrently with exploratory trials, in which the researcher contributes ideas.

In many cases a preliminary needs assessment (taking a few weeks or months) is followed quickly by exploratory experiments. Through monitoring these experiments and discussion at the end of the season, the process of assessing, and reassessing, needs continues. This works if researchers allow themselves enough time for quality interaction with farmers, carry out genuinely exploratory experiments, maintain an open mind on the problems, and do not insist on repeating the early experiments over several seasons in order to obtain 'conclusive' data.

Identifying Where to Work and with Whom

Selecting Farming Systems and Zones

In determining the areas and production systems in which the participatory research will be conducted, both biophysical and socio-economic factors are relevant. Secondary data sources should be consulted, and full use made of whatever information is already available so that duplication is avoided. In many countries maps already exist that identify the various agro-ecological zones: information on the spatial distribution of different livestock species and ethnic groups can be obtained from census data.

Some of the information required may, however, have to be collected by the project through short overview surveys, to enable characterisation of farming systems in a way that is most relevant to the project. Initial characterisation can be modified as further information is collected during the course of a project, and if time is pressing target groups can be developed iteratively, during needs assessment, monitoring and on-farm experimentation.

Pastoralist is defined here as a household in which at least two-thirds of its income comes from sale of livestock and livestock products or livestock-related activities. Thus, their livestock production systems are commercial by definition.

Semi-commercial means that the sale of animals to meet contingencies (i.e. use as liquid assets) and the sale of animals as a profit-making enterprise are both important.

If a project is oriented towards a particular commodity (e.g. buffalo), it may adopt a different approach to targeting from one which is oriented to a particular area or category of farmers. It may, for example, be appropriate to choose an area that is known to be important for the livestock species concerned. In the case of the BAIF/NRI project, the researchers were aware that the main purpose and benefits of goat-keeping vary, and production systems vary accordingly from one area and group to another. Although some of the problems encountered are broadly similar, the most appropriate technologies for addressing them may also vary. Thus, we chose to work in districts that between them represented a range of different production systems.

Initially, the project was only working in Rajasthan (with small and marginal farmers) and Gujarat (with landless or near landless pastoralists). We also wanted to work with landless labour households, in which one or more members were involved in agricultural wage labour. This type of goat-keeper was not very common where we were working initially, so we expanded the project coverage into Maharashtra and Karnataka, where they constituted a larger proportion of the rural population. We also wanted to work more with women, and we knew where there were landless women goat-keepers in Maharashtra.

Multi-locational or multi-production system trials have a further benefit. Scientists gain a clearer picture of production variability in the on-farm trials (from location to location or production system to production system), so they are in a better position to judge the situations and locations (recommendation domains) in which the new technology could be successfully applied (Waters-Bayer, 1989).

Selection of Research Locations

Practical considerations will inevitably limit the choice of specific research locations – 'villages', 'communities', or perhaps a network of local specialists. The further away these locations are from the researchers' base(s), and the greater the distance between the participating farmers, the greater the costs in time and fuel, and the less contact there is likely to be between participants and researchers. Trade-offs may be necessary between the extent to which the locations included are representative, and the resource costs involved. A guiding principle is that well-informed choices are always preferable to the selection, by default, of non-representative situations (e.g. adjacent to research stations, major roads, previous projects, a researcher's home village, etc.).

Where researchers or collaborating organisations have already been working with certain villages for some time, and have developed a good rapport with community members, this may be a strong reason for selecting such villages in preference to others, provided they are reasonably representative of villages in the area concerned. This can save time and resources in that a good rapport with participants already exists, and the project may easily access valuable secondary data about livelihood systems, social and economic composition and problems and priorities.

Selection of type of Livestock

In cases where a project is working in a particular geographical area, and is not tied to any particular livestock species, the researchers will need to decide which livestock species they are going to prioritise. Selection criteria that should be considered are: the potential for research to address livestock-keepers' (not researchers') priority problems associated with the most important type(s) of livestock; where research is likely to produce the greatest benefit, taking account of the seriousness of the problem (e.g. size of mortality rate) and the number of animals experiencing it; and what kind of research is likely to provide the greatest benefit to the poorest groups.

Identifying Participant Livestock-keepers

Options for engaging participants include: (a) volunteering (as individuals or community representatives); (b) delegation of selection to the community; (c) probability sampling (for a discussion of conventional approaches to sampling in agricultural projects); (d) guided purposive selection. Researchers have tended to take a somewhat ad hoc approach, and/or to favour options (a) or (b), on the basis that they are more participatory than (c) and (d).

Approaches (a) and (b) tend to bias the selection, skewing participation away from the poorest, for two reasons. First, within communities power is distributed unevenly and often volunteer or community-nominated participants are male and resource richer. Second, for many of the poorest a prolonged involvement in research activities is not attractive, as they are preoccupied with more pressing livelihood issues. FPR projects need to engage in more systematic selection strategies if they want participants to be generally representative or from particular socio-economic groups. The selection procedure needs to be discussed with the collaborators, and agreement reached on criteria and objectives.

Purposive rather than completely random selection is likely to be the most feasible approach. Purposive selection requires a prior understanding of the socio-economic composition of the village or community and inter-household relations so that farmers' views and reactions can be seen and understood in context; the project should seek to improve its understanding of the local social structure as it progresses.

Techniques such as wealth-ranking and social mapping, if used with skill and sensitivity, can provide the kind of information that is needed initially; and secondary data should be utilised when available. If the relevant information is not available when the initial participant selection process takes place, and volunteer or delegated sampling is used, the project should subsequently check the characteristics of the participants against those of the community as a whole. Additional participants can then be selected if necessary, to make the sample more representative of the target group.

Research objectives are also likely to influence the type of collaborator required. If the research is focused on a particular type of livestock, or a particular kind of livestock production system, then the participants will have to be people who meet these criteria. For example, in a scavenging poultry project working in Tamil Nadu's Namakkal District, three categories of production system were identified, and each category is likely to be represented roughly equally when trials begin.

If the research is testing a new technology the project staff may decide that it is necessary to select willing risk-bearing participants with more resources (e.g. land, labour, equipment) and/or previous positive experience in technology innovation. For programmes that run for a long time, there is a question of whether to continue collaborating with the same small group of livestock keepers, or to change every so often. Generally, this issue has to be looked at in relation to research objectives, and to the importance of maintaining rapport and relations with the community. In practice, it is likely to be expedient to maintain contact with some of the more interested farmers over a period of years, and also allow space for new farmers to join in as others decide to drop out or as new opportunities arise as the experimental programme expands.

Selection of Experimental Animals

The type of animal selected will be determined to a large extent by the nature of the trial. For example, if the experiment is focusing on weight gain in young animals the animals selected will have to be from the relevant age group; whereas, if it were focusing on conception rates they would have to be mature females. Within each of these categories, however, there is scope for varying degrees of precision. This relates to the problem of inter-animal variability. In order to minimise variability the animals should be as similar as possible.

In one of the first trials conducted by the BAIF/NRI project, which focused on young goats, the age spread of the young goats was quite large, creating unnecessary variability and making the use of a standard treatment for all of them questionable. In a similar trial the following year the age of the goats selected was more homogeneous. To minimise inter-animal variability in trials that are focusing on milk production, it may be necessary to select animals that are all of the same lactation, and perhaps ones that are all at a similar point in the lactation cycle.

Another consideration is that where animals in the treatment group belong to different people from those in the control group, the animals in each group should belong to many different owners. Otherwise, the practices of someone owning a large number of animals could become confounded with the comparison between treatment and control groups. For example, in the BAIF/NRI project's first Bhilwara trial, 13 of the 25 goats in the treatment group were owned by one person. Thus, although the treatment group does produced more kids than those in the control group, the difference could have been due to this one goat-keeper having superior goats or feeding practices, rather than to the treatment itself.

Identifying Interventions

In the early stages, the aim of the discussions with farmers is to reach agreement on the research agenda. Deciding on the research agenda is a process which should be based on an adequate understanding of the local farming system, including interactions between various components and enterprises in the system, and who is involved in, decides on and benefits from the various activities. This understanding will help to reduce a long list of possible experiments, to one or two which are most useful and likely to bear fruit. While dialogue between researchers and farmers in this process is essential, dialogue with other knowledgeable researchers may also be vital, in order to avoid duplication and unproductive experimentation. If a consultative mode is used within a national agricultural research system (NARS) setting, further discussions and consultation with other specialists on the extent of the problem and what can be done about it may be required after the needs assessment. Within a community-oriented collaborative mode, this is a joint process between the local people and the researchers.

Whatever the mode, to sustain a credible partnership with farmers and other stakeholders, the probable relevance of possible interventions needs to be gauged through careful study and widespread consultation. If researchers are convinced, but the collaborating farmers are reluctant, it may be worth organising a farmer tour to visit an area where this technology is being practised, or to a research station, before trying to introduce it in an on-farm experiment. The researcher should try and avoid the temptation to tell the local people what to do early on in the discussions. Ideas for interventions may come from any of three general sources:

- members of the local communities
- other livestock-keepers or NR users in the region
- researchers (and extensionists), based on their own organisation's work or the general body of scientific knowledge.

The local people should be encouraged to develop their own ideas initially. It may be useful to discuss ways in which group members have already tried to tackle the problem previously identified, and what effect this had.

Discussions should also screen indigenous technical knowledge and previous experimentation by villagers. Often there are recognised specialists within or near a community, and it may be worth identifying these and inviting them to join in discussions, or making visits to them later for more in-depth discussions. For example, in Rajasthan there are local specialists in animal health care in many villages, called Gunis. In recent years there have been numerous initiatives in different parts of India (by NGOs, KVKs etc.) to document ethnoveterinary knowledge and practices, so it would be wise to check whether there are any relevant studies available in your area or nearby.

Identifying Interventions – Some Examples

Members of the local communities In a feed supplementation trial in Dharwad, Karnataka, the idea for the treatment came from one of the goat-keepers. The treatment was a mixture of sorghum and horsegram. Other livestock-keepers in the region In a de-worming trial in the same district the idea for the treatment came from the practice of another ethnic group from a nearby area, who keep buffaloes. The treatment was the trichomes (hairs) from the pods of a leguminous creeper, mixed with jaggery, a lukewarm sugary solution.

Researchers or extensionists In a supplementation trial in Bhavnagar, Gujarat, the researchers suggested the use of Urea Molasses Granules, which they had recently tested in a pilot project elsewhere in Gujarat. An example from the DAREP project, Kenya Mange had been identified by goat-keepers as a major problem. Farmers considered commercial products for treating it to be too expensive, and had started looking for locally available alternatives.

The project staff sought to identify alternatives by holding a number of group discussions and by visiting local herbalists, and came up with a list of eight local concoctions already being tried by farmers. The list was further screened through discussion with farmers, and a trial was designed comparing three of the local treatments with two of the recommended commercial ones and a herbal treatment of Neem solution.

Understanding the Nature of the Problem

This issue is discussed in the companion volume to this one, where tools for determining priority needs are described (Conroy, 2001). These include: (a) participatory problem tree analysis, which livestock-keepers can use to separate the core problem from its causes and effects; and (b) the participatory herd history method, which is a quick method for generating reasonably reliable baseline data on kidding rates, mortality rates etc. Understanding the nature of a problem correctly requires researchers to know what the objectives of the livestock-keepers are.

Sometimes the livestock-keepers (or the researchers) may not know, or be sure of, the nature of the problem; but they may have an informed hunch as to the cause. In these situations it is essential for both parties to discuss and explain their ideas and hypotheses before selecting treatments. One common problem that is sometimes poorly understood by goat-keepers is that of high worm burdens, which can result in high mortality rates in kids due to a combination of factors. A high worm burden in the mother may lead to a dramatic reduction in milk production, thus weakening the kid; and if the kid itself then becomes infected it will have a poor chance of survival (Peacock, 1996). However, the goat-keepers may not realise that helminths are causing the death of their kids, nor understand the life cycle of helminths and their effect on the goat's physiology. If so, the researchers will need to educate the livestock-keepers on "the underlying principles of the technology effect" (Mason et al., 1999).

This is similar to the situation with integrated pest management of crops, where 'farmer field schools' have been developed as an educational tool (Ooi, 1998). The high importance livestock-keepers attach to visible problems may also mean that they prefer to take a curative approach (i.e. respond to the problem when it appears) rather than a preventative one. This was found to be the case in one de-worming experiment on cattle and sheep (Mulira et al, 1999).

Determining the Size of the Treatment

This should be based on a combination of technical and cost factors. For example, the quantity of a supplement or a drug should obviously be related to the size of the animal. However, the cost of the treatment should also be taken into account. Livestock scientists may be inclined to choose the quantity that will have the biggest physical effect, but that is not necessarily the most profitable size of treatment. Furthermore, poor livestock-keepers are usually very short of cash, and therefore have to ration what they have among different items of expenditure: thus, they may not be prepared to incur the full cost of what the scientist considers to be the ideal quantity. If enough animals are available for the trial it may be possible to test two or more different quantities of a treatment.

It can be useful to make a simple appraisal of the likely profitability of the proposed treatment, particularly those that involve the purchase of an input. If the treatment has been clearly defined the cost is usually easy to calculate and predict. The size of the benefits is uncertain, of course, but comparing potential benefits with a range of "guesstimates" of the benefits will give some indication of whether the technology concerned is worth investigating. If, as happened with one of the proposed treatments in the BAIF/NRI project, the expected benefits are about the same as the predicted costs it would be advisable to reconsider the proposed trial. This test can be a useful check on ideas coming from researchers with a technical background, who may not be used to thinking in terms of profitability.

There are several different tools for analysing the profitability of treatments or interventions. They have differing degrees of sophistication and slightly different uses, and each has its advantages and disadvantages. The ones described here are among the simplest available, and should be adequate for most types of on-farm trials. For information about other techniques the reader is referred to Amir and Knipscheer (1989) and to standard text books in agricultural economics.

Partial-budget analysis is a simple method for estimating the profitability of a treatment, which involves the tabulation of expected gains (benefits) and losses (costs) due to a relatively minor change in farming methods. (The term partial indicates that the change only occurs in one component of the farm.) It is widely used to determine the profitability of a single intervention. An example of a partial budget analysis from the BAIF/NRI project. It assumes that the goat-keepers' primary objective is to maximise the number of kids produced, so that they can either be sold or retained to maintain or increase the herd size. The value attributed to a kid under the cautious assumptions is the value it would fetch if sold shortly after birth; whereas that under optimistic assumptions is the price it would fetch at about six months of age (say Rs 800), minus the cost of rearing it during that six months (say Rs 300).

Sometimes it is more appropriate to develop partial budgets for a whole herd than for a single animal, as in this example. Profitability has been estimated per 10 does, rather than per doe, because the treatment does not affect every individual: some does will produce more kids than they would have without the treatment (e.g. one instead of none, or twins instead of one), but others will be unaffected. The analysis shows that the treatment should be profitable, even on the basis of cautious assumptions.

Costs of items that are not traded can be difficult to quantify. In the earlier example, there is a small market for tree pods, so the market price has been used in the analysis. However, the goat-keepers were collecting the pods themselves, so it would have been more appropriate to estimate the labour cost of doing so; but since there is no wage payment involved, determining the labour cost is not entirely straightforward. To do this it would have been necessary to estimate the opportunity cost of labour: that is, the cost of collecting the pods would be equivalent to the value of the best employment alternative that the goat-keeper had foregone so that (s)he could collect them.

For example, if (s)he had to give up two days of wage employment at a rate of Rs 40 per day, then the labour cost of collecting them would be Rs 80. In rural areas, the opportunity cost of labour tends to vary considerably over the course of a year: it depends partly on how much work is available, with wage rates being higher during busy periods. As it happens, the pods are collected at a time of the year when there is little work available, so the opportunity cost would be quite low in this case. If there was no alternative work available, then the opportunity costs would be zero.

Break-even analysis In partial budgets there are always a few key factors that affect the balance of gains and losses. Break-even analysis determines the level at which the gains and losses are equal: this level is known as the break-even point. Generally, break-even analysis is done by manipulating the most uncertain key factor.

When used in association with partial budgeting it can be used as a measure of risk, indicating at what value of a critical factor a new technology is expected to no longer be profitable to a farmer. Break-even analysis can be applied to the tree pods example, where the number of additional kids is considered to be the most uncertain factor. Thus, the researcher (and livestock-keeper) wants to know, given the estimated costs of the treatment, what is the minimum number of extra kids required to enable the owner to break even.

Benefit: cost ratio is simply the ratio of benefits to costs. It is normally calculated in financial analyses that take account of the flows of costs and benefits over many years (the duration of the activity – e.g. a commercial poultry unit), and that discount future costs and benefits.

Thus, in a financial analysis it would be the ratio of discounted benefits to discounted costs. For simple technologies that have fairly immediate effects there is no need to do a discounted cash flow, but it can still be very useful to calculate benefit: cost ratios. They show how much extra benefit the livestock-keeper is likely to get for each unit of cost incurred – in other words they give an indication of the likely size of the return for a given level of investment. If farmers behave rationally then they will be most likely to adopt those technologies with the highest returns, all other things being equal. The benefit: cost ratios have been calculated for the tree pods technolog. However, where significant cash expenditure is involved, particularly in 'one go', they may not be able to access enough money to pay for the all of the input; or they may avoid spending money if they believe there is a high degree of risk involved.

Standardising the treatment

For trials that are seeking to quantify the size of the benefit brought about by a particular treatment it is important to standardise that treatment across all the animals in the treatment group. (This will not be an objective in some trials, which may be more interested in seeing how the livestock-keepers modify a treatment developed by researchers.) For trials in which the treatments are feed supplements it is desirable to provide participants with a measure (e.g. a beaker) that corresponds with the quantity of the treatment.

For treatments that are more complicated, such as ones requiring the processing or mixing of ingredients, it may be necessary for the researcher to provide the treatment. For example, in the trial to address mange in goats, the local concoctions were supplied and prepared by the researcher (Sutherland and Kang'ara, 2000).

When to Experiment

Sometimes it is necessary to consider whether an experiment is required. If you know of a technology that has proved effective in similar situations elsewhere, it may be appropriate to introduce it without following a rigorous process of experimental planning, implementation and evaluation? While everything new to an area may be seen, in certain senses, as an experiment by farmers, not everything new may need a formal experimental design. Through discussion with farmers, especially any with some experience of laying out trials, it may be possible to classify those interventions which require a formal experimental design, and others which can be introduced or tested in a less formal and less resource intensive way.

Identifying Experimental Hypotheses

Once a treatment has been chosen for a formal experiment, it needs to be formulated in terms of a precise hypothesis. It is important to involve the livestock-keepers in the process, and it need not be difficult to do so. The format (Veldhuizen et al., 1997) illustrated by the following example can be used.

If... I give a dewormer to my does in late pregnancy and at time of kidding

Then... Their kids are less likely to die young

Because... 1.Their mothers will produce more milk and thereby make them stronger, 2. Their mothers will not transmit worms to them through their faeces.

Veldhuizen et al. (1997) stress the importance of hypothesis formulation in the context of the collegiate mode as follows:

"This is a crucial step in the dialogue between farmers and outsiders. It helps the researchers and collaborators to define more precisely what they want to try out and why, and enables them to analyse more clearly the results of the trial. It is a planning, monitoring and evaluation tool. It helps both parties to understand each other's logic better. It provides an opportunity to check the reasons for the problem and prevents jumping to conclusions about possible solutions".

How Many Livestock-Keepers and Animals should be Involved in an Experiment?

In considering this question it is necessary to bear in mind a number of factors.

1. The cost and logistics of meeting participants and collecting data.
2. The quality of interactions with participants versus the quantity.
3. The time required to process and analyse monitoring data.
4. The minimum number of sample units (fields, animals) required to be able to draw general conclusions and for statistical analysis (if required).

Trade-offs may be required between factors 1, 2 and 3, on the one hand, and the fourth factor. Poor judgement may lead to serious errors of one kind or another. Too few sample units (point 4) may mean that the sample is not representative or that firm conclusions cannot be drawn from the data. Too many sample units, on the other hand, may mean that the researchers are overwhelmed by the amount of data that they need to collect and analyse, and that this results in errors of recording, coding, data entry or processing. In addition, limited interaction with livestock-keepers may result in them not identifying strongly with the trials, and not understanding or correctly following the trial design.

One general guideline is that is best to avoid being over-ambitious to begin with – mistakes in small trials are less costly than mistakes in large trials. Where statistical analysis is not involved, there are no hard and fast rules for determining sample size. One suggestion, however, is that in the initial phase, if fieldworkers are learning to work in a participatory mode with farmers, the number of experimenting farmers per fieldworker. One factor that should be taken into account when determining the number of animals required is the probability of some animals leaving the herd during the course of the trial - due to sale or death, for example. The necessary number of animals should be added to the minimum to allow for this. Researchers can, of course, try to persuade livestock owners not to sell their animals until the trial has finished, but in a participatory trial this possibility cannot be ruled out.

How many Treatments?

Since farmers' situations may differ considerably, it will probably be desirable to test several interventions through experimentation. However, more interventions will generally mean that more participants and resources are needed, if statistical analysis is to be done. There may have to be trade-offs between the number of experiments required by the range of situations in which the project is operating. From the point of view of data analysis, it is better to collect meaningful data on a few experiments, or on one experiment with a modest number (1–2) of treatments, than inadequate data on a larger number. However, from the individual farmers' point of view, it may be preferable to have a large number of simple (i.e. with and without or before and after type treatments) experiments which carry fairly low risk and may be superimposed on existing farming practices. (An example of a 6-treatment trial is given in Box 4.1.) There is room for negotiation between researchers and livestock-keepers on this topic.

Experimental Design

Experiments seek to test hypotheses by comparing treatments: there must, therefore, be some basis for comparison. In principle, the animals on which the treatment or intervention is to be tested can be compared: before and after the intervention; or with animals on which the treatment was not tested (a with and without comparison); or with animals on which a different treatment was tested.

Each method has its advantages and disadvantages, which are discussed in the following sections. Which is more appropriate may depend to some extent on the nature of the experiment.

The 'before and after' Method

The before and after method can be used in such a way that the comparison is made for the same animal. When this method is used "a sequence of experimental treatments is randomly allocated to each farm animal, and there are several different treatment sequences" (SSC, 2000). Each animal is given a treatment for a fixed period of time, and then changes over to another treatment. For example, a cow's milk yield can be measured before feed supplement 'X' is given and after, and then before and after feed supplement 'Y', and so on; and the effect of each supplement can be determined in this way.

When used in this way the before and after method has the major advantage of excluding inter-animal (and inter-farm) variation as a factor influencing the parameter being measured: this kind of variation is usually larger than within animal variation, group of four cows that were receiving the supplement and another group that were not, differences between the two groups could have been influenced by several other factors, such as general differences in diet between them or differences in the stage of lactation or parity. However, this method also has potential disadvantages. First, any changes that occur may be influenced by extraneous factors that have changed over time, such as rainfall or forage availability, and these could be more important than the effect of the treatment. It is not well-suited, therefore, to experiments lasting several months, in which the before and after data are from different years or different seasons. An example of inter-annual differences is given in Box 5.1. Seasonal variations may also be marked, for example growth rates or milk yields between dry seasons and rainy seasons.

Alternative Poultry Production Systems

Defining Production Systems

Alternative production systems are defined in some countries in order to assist in marketing so that consumers will understand how the birds were raised. For example, in addition to maximum stocking densities for indoor and outdoor areas, the type of breed and feed may be specified. There are few definitions in the U.S.

The following classifications and general definitions constitute accepted terminology among both conventional and alternative poultry producers.

Layers

Poultry are mainly raised in the following ways for egg production:

- Cages

- Cage-free or "barn"
- Free-range.

In the U.S., these are not regulatory terms. In contrast, in the European Union, they are "special marketing terms" with legal standards and definitions.

Cages. Indoor production is used by the large-scale conventional industry and is usually climate-controlled. Large houses allow intensive production and a high level of automation of feeders, waterers, etc. Layers are housed in cages in conventional indoor production, usually tiers of adjacent cages or "battery cages." Most U.S. producers provide 67 to 86 square inches per hen. However, many consumers are concerned about the small amount of space in battery cages and the fact that birds cannot perform natural behaviours such as nesting, perching, and scratching. In the European Union (EU), battery cages will be banned by 2012. Only enriched cages can be installed, which provide more space (at least 116 square inches), as well as nesting, perch, and scratching areas.

Cage-free. Cage-free layers are raised loose on a floor, which is generally covered with litter to absorb manure and allow birds to scratch. Eggs from cage-free layers are sometimes called "cage-free" or "barn eggs."

Maximum stocking density should be no more than one bird per 1.5 square feet, but can be increased up to 1.2 square feet with slats and multiple levels of flooring on litter. Slatted floors permit a higher stocking density since birds roost on the slats at night, droppings fall into a pit below, and less manure accumulates in the litter. However, some programs limit the amount of slats that can be used in the house in order to ensure birds have sufficient solid floors with litter for scratching. In the EU, the "Barn" system requires 1.2 square feet per hen and 39 square inches of litter, along with perch space.

Aviaries are specialty, multi-tiered buildings for cage-free layers that provide several levels of flooring and use vertical space (perches and platforms) to allow birds to jump to different levels. Aviaries can actually maintain a high density of hens.

Free-range. The U.S. Department of Agriculture (USDA) does not currently have specific regulatory definitions for "free-range," although the term is allowed on labels under certain circumstances. When applying for label approval, the producer must submit a brief description of the housing, which the USDA reviews to determine that poultry have access to the outdoors for at least half their lives. In the EU, the definition for "free-range eggs" requires outdoor access with a maximum outdoor stocking density of 1 hen per 43 square feet (the equivalent of 1,000 hens per acre), and also requires the use of the housing in the "Barn" system described above.

Meat Chickens

Meat chickens are generally raised in two ways:

- Indoors on litter
- Free-range.

Indoor. The large-scale conventional industry raises broilers in houses. They are raised on the floor on litter, usually at a stocking density between 6.5 to 8.5 pounds per square feet, which is less than 1 square foot per bird. Sometimes birds are provided with additional space and marketed on the basis of having extensive space indoors.

Free-range. As mentioned, the USDA does not have specific definitions for free-range. In contrast, the EU defines "Free-Range" and "Traditional Free-Range." These definitions specify maximum indoor and outdoor stocking density, the type of feed (70 percent cereal at finishing), the minimum slaughter age, and the amount of pophole space to encourage birds to go outside.

"Free-range" limits indoor stocking density to 5.5 pounds per square feet and outdoor to 10.8 square feet per chicken (4,033 chickens per acre). "Traditional Free-Range" is basically the French Label Rouge standards and requires more space. Stocking density is limited to 5.1 pounds per square foot indoors and 21.6 square feet per chicken outdoors (2,026 chickens/acre). It also requires the use of a slow-growing breed and limits the number of birds per house and the number of houses on a farm.

Outdoor access is also required in most organic programs in addition to the use of organic feeds, preventive health care practices, and a prohibition on the use of antibiotics. The USDA National Organic Program requires outdoor access but does not specify stocking density or flock size. Instead, the standards are more descriptive, requiring fresh air, direct sunlight, the opportunity to express natural behaviours and exercise. In contrast, EU organic legislation specifies maximum stocking densities both indoors and outdoors.

Stocking density/flock size. In general, stocking density is related to bird welfare and carcass quality. Outdoors, stocking density is used to maintain vegetation and reduce pathogens and excess nutrients. Programs have specific stocking densities, but as a general rule-of-thumb indoor maximum stocking density for meat chickens should not be more than 1 bird per square foot or about 6 pounds per square foot. For the outdoor area, the Soil Association recommends no more than 1,000 meat chickens per acre or 400 hens per acre on pasture that is rotated. Cage-free layers should have at least 1.5 square feet per hen indoors on litter. Meat chickens can be stocked at a higher rate than layers, because meat chickens are around for a shorter time.

Many programs limit the size of the flock or number of birds in a house. The Soil Association recommends that flocks be kept to 500 birds or less but allows flocks of 2,000 hens or 1,000 meat chickens.

Free-Range Systems

Free-range systems vary widely, from large fixed houses with yards to small portable houses that are moved regularly. The main free-range production systems are

- Fixed houses
- Portable houses
- Pasture pens
- Integrated systems.

Fixed Houses

Advantages to a fixed house include the relative ease to provide electricity to power automated equipment. The house can also be large in size because it is not moved. Fixed houses open to yards to provide outdoor access, and birds are usually closed in the house at night. This type of production is done by most large free-range poultry companies in the U.S. and is sometimes called "semi-intensive." On a small scale, it is the familiar chicken coop and yard.

A major disadvantage of this system is that unless the birds are rotated from one yard to another, access to the same ground is continual. When birds stay on the same ground, they damage the vegetation and turn the yard into dirt or mud. Turf damage may be especially bad right around the house, and birds may track mud into the house and dirty the eggs. Excessive pathogens and nutrients can build up in the yard, contributing to disease and pollution. A fixed house and yard may work in very dry climates, such as parts of the western U.S., where biological activity is low and pathogens do not survive, or with a low density of birds. (Young pullets or broilers are more susceptible to pathogens than older birds like layers.) Otherwise, it is critical to rest or rotate pastures to prevent these problems. If the house is fixed, a rotation should be used to rest the pasture.

One way to help rest or rotate pasture is "double yarding"—subdividing the yard in two with a fence and rotating the flock between the yards. The Soil Association recommends dividing the yard into at least four paddocks and planting trees or using shelters at least 30 to 60 feet away to draw birds from the area around the house, which is the most heavily used. In the Soil Association program, the pasture must be rested for nine months after every laying flock; for meat chickens, the pasture should be rested at least two months per year, plus one year in every three years. A layer of mulch or gravel around the house helps reduce mud outside the popholes and keeps birds from tracking mud into the house. Slats on either side of the popholes also help clean feet. Fencing may be permanent or temporary. Temporary fencing such as electronet fencing can be used and is easily moved to provide more yards.

Ideally, yards are covered with vegetation but sometimes are simply dirt scratching or exercise areas. Ground coverings such as gravel, straw, mulch, or sand are preferable to a dirt lot and help reduce mud. The area may even be completely or partially covered with a roof, making a veranda. Screens may be used to enclose the birds in a "winter garden" or curtains can be used. This allows access to sunlight and fresh air while protecting birds from weather and wildlife. In some cases, a screened porch with a mesh floor allows manure to pass through so it can be collected and removed.

Portable Houses

Mobile houses are necessarily small since they are built to be moved regularly to a new location, usually with a tractor, pickup, all-terrain vehicle, or draft animal. Portable housing ranges from crude shelters to well-constructed, insulated houses. Houses may have wheels or skids and are moved every few days or less frequently. Pastures usually have a perimeter fence to contain cattle or other livestock and deter predators. Generally poultry producers confine birds at night to protect them from nocturnal predators and the elements.

Moving the house at least once a week prevents the pasture underneath the house from dying. It recovers in about a month, depending on the climate. If the house stays in one location for longer than one week, the forage plants under the house may need to be re-seeded and it may take a couple of years to re-establish new pasture in that spot.

If a layer house has wheels (an "eggmobile"), it is usually moved often. Virginia producer Joel Salatin popularized the use of eggmobiles in the U.S. His layer houses moved through pasture every three to four days following grazing cattle. Old mobile-home trailer frames and old campers have also been used for eggmobiles. According to Salatin, significant acreage (50 acres) is needed in order to move the birds far enough that they do not return to the previous spot or identify a favorite spot such as a garden. He moves his eggmobile every few days.

Skid housing is generally moved less often than wheeled housing, because it can be hard to drag. If wooden skids are used in organic production, they should not be pressure-treated with copper-chromium arsenate or other prohibited materials. Metal pipes are another option and are smoother to drag than wood.

The use of small shelters on skids is described in *Free-Range Poultry Production and Marketing* by Herman Beck-Chenoweth. His shelters have a roof and a wood floor covered with litter but only chicken wire instead of walls. He moves the shelters to fresh pasture every four weeks to prevent turf damage and maintains at least 100 feet between each house to keep flocks separate. Beck-Chenoweth's system is described on his Web site. Many attractive designs for small poultry houses used for free-range production in the United Kingdom are shown on the Forsham Cottage Arks Web site. Andy Lee pioneered the use of electronet for poultry in the U.S., using the term "day-range," in his book Day Range Poultry. The use of portable electrified netting as fencing allows bird activity to be placed where desired, reduces predation, and is easy to move. A 165-foot roll of netting encloses an area roughly 40 feet by 40 feet.

The electronet may completely encircle the house, or it can be looped off one section of the house. Temporary fencing makes many configurations of paddocks in conjunction with a mobile house. The electronet should be moved before the grass starts growing into the netting, which will short it out and make it hard to move. Putting the housing in the middle of the enclosure will reduce birds flying out.

Salatin has an egg production system using net fencing that he calls the "feathernet." Salatin moves his "feathernet" house and netting to a new site every three days, making a figure-eight with two 450-foot circles of netting (using three rolls each), so he can move the housing into a new circle without letting birds escape.

Housing is sometimes designed to remain open at night in order to provide longer hours of outdoor access and eliminate the tasks of opening and closing doors. Birds forage actively at dusk. EU legislation allows a greater indoor stocking density if the house remains open at night. Flaps over the exits can deter predators such as owls from entering the house and help maintain temperatures in the house, and electronetting deters ground predators.

The housing design may impact portability. Portable houses usually have litter-covered floors to protect birds and keep them off wet ground. Floors increase the structural stability to keep the house from pulling apart when moved. However, houses without floors are less expensive to build and reduce potential habitat for rodents. If a portable house has no floor, it is important to keep rain water from seeping into the house through the ground. A trench above the house on a hillside can divert water flow from heavy rain, or ideally, the house is placed on ground that is higher than the surrounding ground or on a constructed pad of earth.

Colony Production

Colony production is a management system for free-range egg production described by Oregon producer Robert Plamondon on his Web site. Colony production uses many small houses along with a common nest house and feed area. It is based on a system that was popular in the early 1900s in the California poultry industry centered near Petaluma. It was designed to reduce labour in gathering eggs from small portable houses on range. "Colonies" of small roosting houses share a common nesting house and feed area. Plamondon moves his houses every few weeks to a new spot about 100 feet away. The producer visits only one house, the nest house, to collect eggs. The producer collects the eggs onto flats for pickup by truck or ATV (or, in the past, horse). According to Plamondon, when you have more eggs than you can carry by hand, you need a nest house. Plamondon's roost houses do not use floors or litter, although nesting houses have litter to clean birds' feet. When he moves the house, there is a 2 to 4-inch layer of manure left which he scrapes with his tractor and incorporates into the pasture. The houses should be kept 300 feet from barns, garages, and other places where you don't want birds to roost. He keeps 50 hens in each small house, and his colonies consist of 200 hens each.

Pasture Pens

Pasture pens are small floorless pens that are moved daily, usually by hand, to fresh pasture. These inexpensive shelters are set in a pasture, on a lawn, or in a garden. They are a favorite of small-scale producers due to their low cost and flexibility.

Pasture pens were popularized in the U.S. by Joel Salatin of Virginia. Salatin wrote a book called Pastured Poultry Profits which described his system to feed, provide care, market, etc. The pens have been so popular with small farmers that Salatin is credited with starting a "pastured poultry" movement in the U.S. He uses a 10-foot by 12-foot by 2-foot wooden pen. The top of the pen is flat, and partially covered by roofing. He stocks at a maximum density of about 1.5 birds per square foot for meat chickens. Producers following his book often have a high rate of success, but moving the pens daily is very labour-intensive.

No litter is used so birds can forage for grass and the pen is moved daily. The daily moves control coccidiosis, a parasitic disease that occurs if birds are in contact with their own manure. Salatin advises to return to the same plot of land only once per year.

Since the field pen is only a shelter, the birds are usually raised only in warm weather. On rainy days, producers put hay bedding on the ground under the covered section of the pen to help keep birds dry. One producer uses a "sled" under the pen; a good way to get the birds off the ground in rainy weather.

The sled is smaller in dimensions than the pen. She lifts up one end of the pen and pushes the sled in. She uses a rectangle frame that is covered with hardware cloth so droppings still fall through to the ground. Although the birds are not free to roam in a pasture pen, they still have the advantages of fresh air and sunlight. However, the confined space of the pen is a welfare concern. Larger pens are needed for turkeys because of their wing spans.

A field pen generally provides good predator control, but some predators (namely raccoons) can grab chickens through the wire. If there are low spots in the pasture, holes between the bottom of the pen and the ground need to be stopped up with scraps of wood or other materials. Some producers run an electric wire around the pen area a few inches off the ground to keep predators away or surround the pens with electronet fencing. Some producers use a combination of field pens and fencing, opening up field pens within a yard enclosed by electronet; the birds range in the entire enclosure during the day.

The pens are moved manually by putting a dolly on one end and lifting by a handle on the other. Birds learn to walk along with the pen as it is dragged across the field, but occasionally a bird will escape or be injured during a move.

The basic pen design can be modified in many ways. Plastic (PVC) pipe and rebar have been used in place of wood to lighten the structure. However, in areas with strong winds, light pens need to be staked down. Some producers peak the roof to allow more heat to escape, to keep rainwater from pooling, or to keep goats off. Skids or wheels can be mounted on the pen instead of a dolly to move the pen. The field pen can also be adapted for egg production by adding nestboxes. There are many other field pen designs that are flat, peaked, domed, or hooped. Building materials include wood, PVC, rebar, electrical conduit, and bamboo. Plamondon has a hoop design on his web site that uses cattle panels. In Europe, there are some fancy—and expensive—pens. Some have attached housing. These European "ark" houses are available through U.S. distributors, but there are also some similar pens in the U.S. The Henspa and the Eglu are available for urban poultry production.

A booklet called Pastured Poultry, developed by the National Center for Appropriate Technology (NCAT) for Heifer International, is available from ATTRA. It presents case studies from a Heifer project, describing the experiences of 19 producers in the South who each raised a batch of pastured poultry in field pens, and includes sections on mortality, weather, pen construction, economic analysis, and more.

Integrated Systems

Diversified farming is an important part of sustainable agriculture, and poultry can be integrated with livestock, crop, and vegetable production in "permaculture" systems that integrate principles of natural systems with agriculture. Some organic programs require that at least 50 percent of the poultry feed come from the farm where the birds are raised—or a nearby farm— in order to keep the nutrients cycling in the same region. Diversified farms allow nutrients to be recycled between plants and animals.

Diversified systems focus on services that poultry can provide, such as fertilization, tillage, and insect and weed control rather than only meat production. A permaculture concept called "stacking" combines several enterprises on the same piece of ground. Various species of domestic animals can be raised together to complement each other, creating mutually beneficial relationships.

Disease cycles can be broken when the same species does not occupy the same site all the time. For example, sheep on pasture can clean up parasites that affect poultry. However, poultry species (chicken, turkeys, ducks, etc.) should not be mixed, nor should flocks of different ages comingle. Some species are "carriers" of disease that can affect other species and older birds can also carry disease to younger birds. Salatin has a hoophouse mainly to overwinter small livestock. Rabbits are kept in hanging cages with chickens on the floor. Worm beds are kept underneath the rabbit cages and are covered with wire to keep chickens out. In the spring, after the animals are removed, the hoophouse is used for early vegetable production. Salatin uses double layers on the hoophouse—a shade cloth and a clear tarp. The shade cloth can be removed to capture solar energy during winter. Salatin makes his buildings multipurpose in order to rotate species in them.

Poultry may share pasture with cattle, sheep, and goats. The pasture soil is improved by the poultry's rich manure which helps revive fertility on a farm. Several species of animals may be grazed simultaneously, or the grazing may be staggered to allow only one species at a time in the paddock. Multispecies grazing can aid in protecting poultry from predators that respect large animals. Sheep in particular can help manage the forage for poultry. However, cattle and goats may disturb poultry housing and feed. Cattle may not have experience with poultry netting and may bring it down by accident. Exclude ruminants from poultry feeding areas to prevent foundering from overeating grains. Also, poultry feed may have additives not appropriate for ruminants. Many cattle producers keep poultry to scratch apart larva-harbouring dung pats, which helps reduce fly and parasite problems on the cattle. Joel Salatin has said he would keep layers in his cattle pastures even if there were no eggs—just for the health benefits to the cattle.

Chicken Tractor

Some vegetable growers insist that in order to build a sustainable system, livestock or their manure must be incorporated into the farm for fertility.

A "chicken tractor" is a way to integrate poultry production with vegetable production. Andy Lee described his system in a popular book called *Chicken Tractor*. Birds are kept in small pens in a garden to provide fertility, tillage, and insect control. Lee uses a small floorless pen so the birds can forage and scratch. The pen is covered with wire and usually has a covered top or a small attached house. The pen is moved daily on fallow beds to add fertility and increase garden yields. The chickens also weed and till the beds and help control insects. Garden wastes are useful feed supplements. In addition to rotating the pen daily to a fresh spot, Lee suggests keeping the pen in one spot and adding fresh straw bedding daily to create a raised garden bed. Moving the pen after one month will leave a sheet-mulch on top of the beds to kill grass and weeds. According to producer Jean Nick, heavy broilers don't really till the soil. "They just poop and stomp on it." Layers are better at clearing weeds and bulbs and scratching the ground.

Scientific studies have examined the impact of poultry on fertility, integrating birds with vegetable and forage production. Jim McNitt, PhD, at Southern University and University of Illinois graduate student Ben Lubchansky have conducted such examinations.

Poultry can also be kept in gardens, fenced with portable electronetting. Chickens help prepare the ground for vegetable planting by tilling. After harvest, birds clean crop residues in market gardens in the fall— turkeys are especially useful for this purpose. According to Andy Lee, "from October through Thanksgiving the turkeys can clean every bit of weeds and spent plants from the garden and leave a rich load of manure behind." "Fold" houses in the United Kingdom allow flocks of chickens to help glean fields after crops are harvested. Chickens are not generally appropriate for a producing garden, because they scratch up seeds or eat crops. According to Vermont producer Walter Jefferies, "I don't let them in early in the season when the seedlings are getting started or late in the year when they'll peck ripe veggies. Chickens, guineas, and ducks all work with some plants such as potatoes, corn, tomatoes at the right states."

Poultry can also be kept in vineyards and orchards. Researchers at Michigan State University studied the use of chickens and geese in apple orchards. Chickens were found to control insect pests while geese aided in weed control. Poultry are kept in some California vineyards. Producers Greg and Jeff Kuntz in Iowa have raised poultry in a multicropping system: a vineyard system with sweet corn and poultry. The brothers use the corn and poultry as a way to add value to the vineyard while the grapes are being established. Young vines are encased in grow tubes that protect them from poultry and that can be removed later. If poultry are kept in areas with growing crops that are certified organic, measures must be taken to ensure that the crop does not become contaminated with droppings to the extent that it presents a food safety hazard. USDA organic standards require that raw (uncomposted) manure be incorporated 120 days before harvest of a crop whose edible portion has contact with the soil (and 90 days for crops without direct contact). While there is an exception for grazing animals, it is still important to take appropriate measures to prevent the presence of manure on crops at harvest time. Other benefits offered by poultry:

- Some people keep chickens for tick control on their land.
- Turkeys were used in the past for insect control in crops, such as in tobacco during U.S. colonial times. Turkeys are more aggressive foragers than chickens.
- Weeder geese were used on a large scale in California in the '50s to weed cotton fields before the widespread use of herbicides. Geese have a strong preference for young grasses and have been used successfully to weed crops such as strawberries, potatoes, and onions.
- Ducks have been used to control aquatic plants in ponds, especially duckweed and pondweed. Muscovy ducks have been used for fly control on dairy farms. Ducks and geese provide insect, snail, and slug control.
- Guineafowl, considered luxury food in Europe, are good foragers, controlling insects in pastures and gardens. Because of the noisy calls they sound when alarmed, guinea fowl and geese can also act as "watchdogs."
- ATTRA can provide more information on turkeys, geese, weeder geese, ducks, guinea fowl, and gamebirds upon request.

Application of Outdoor Poultry Production to International Development Work

The production systems described in this publication are useful not only for alternative poultry production in the U.S. and other developed countries, but also for developing countries, where production systems rely heavily on integration with other farm activities. In many villages, poultry are kept loose in order to scavenge food waste. While large-scale indoor production system is used in many countries around the world, small-scale systems remain important. For example, small-scale growers produce 30 percent of the poultry in China.

Structure of Poultry Industry and Disease Control

The basic structure of the poultry in-dustry can be regarded as a Breeding Pyramid. The apex of the pyramid is occupied by Elite (Primary) Breeders, Great Grand Parents (GGPs) and Grand Parents (GPs), followed by Parents in the middle and commercials at the bottom of the pyramid (Hunton, 1993). The major genetic manipulations take place at elite and GGPs level. This involves careful recording of the main production parameters and also, selection for desired traints. At this level intensive health monitoring is essential, so that diseases which can be vertically transmitted or which can depress performance are excluded.

At this stage, the pathogens detrimental to the production and health of birds are to be eradicated from the elite breeders to GGPs, GPs and Parents can produce devastating results in the commercial poultry industry as hundreds or thousands of commercial birds are descended from each bird of the elite breeder flocks. So the disease status, immune system and genotype of elite flocks have potentially profound effect on the health of commercial flocks. The need to protect elite and GGPs from catastrophic and vertically transmitted diseases dictates high levels of biosecurity.

Concept of Health and Disease

Health is one of the most important factors determining profitability of all systems of poultry production. It is defined as a state of being that is not merely absence of disease, but a state of complete physical, mental and social well being of poultry. Health and disease lie along a continuum and there is no single cut-off point. The lowest point on the heath disease spectrum is death and the highest point corresponds to perfect health. It is thus obvious that health fluctuates within a range of optimum well being to various levels of dysfunction, including the state of total dysfunction, i.e. death.

Health and Disease Spectrum:

1. Perfect Health
2. Better Health
3. Freedom from Disease
4. Unrecognized Disease
5. Mild Disease
6. Severe Disease
7. Death.

Health status not only influences the average performance of a flock but also its variability. The perfect health implies that a bird should be able to express as completely as possible the potentialities of its genetic heritage. This is possible only when a bird is allowed to live is healthy relationship with its environment an environment that transforms genetic potentialities into phenotypic realities.

The disease is Just the opposite of health i.e., any deviation from normal functioning or state of complete physical or mental wellbeing. The term "disease" literally means "without ease" (uneasiness) disease, the opposite of ease when something is wrong with body function. Disease represents a restraint to reproductive potential and yield, feed conversion efficiency and product quality. The causative factors of disease may be classified as Agent (Pathogen), Host (Bird) and Environment (Jordan, 1996). These three factors are referred to epidemiological triad. The mere presence of agent, host and favourable environmental factors is not sufficient to start the disease in bird. What is required is an interaction of these three factors to initiate disease process in the bird. The agent, host and environment operating in combination, determines not only the onset of disease which may range from a single case to epidemics but also the distribution of disease in the flock.

Agent Factors:

1. *Biologic agents:* The biologic agent may include viruses, Chlamydia, rickettdiae, mycophasmas, bacteria, fungi and para-sites, These agents exhibit certain biological properties viz., a. Antigenic variation, b Variation En virulence, c. Variation in immunogenic, d. Infectivity (numbers necessary to establish infection in the host), e. Tropism, f. Survival within and outside the host and g. Resistance/Susceptibility to antimicrobial and disinfectants.
2. Nutritional agents
3. Physical agents
4. Chemical agents
5. Mechanicalagents.

Host Factors (Intrinsic) :

1. Age
2. Gender
3. Immune status
4. Stress (associated with point of lay or social stress).

Environment factors (Extrinsic):

1. Climate
   a) Season
   b) Ambient Temperature
   c) Humidity

2. Air Quality
   a) Dust
   b) Gases like NHs, C02. And CH4
   c) Free radicals.
3. Housing
   a) Direction
   b) Ventilation.
4. Lighting
   a) Intensity
   b) Colour
   c) Duration
5. Management
   a) Floor/Cage rearing
   b) All-in-all out
   c) Multiple ages
   d) Vaccination & Medication
   e) Debarking, Dubbing & Sexing
   f) Handling
   g) Transportation
   h) Hygiene and disinfection
   i) Disposal of dead birds, litter and hatchery waste.
6. Nutrition
   a) Deficiency or excess of nutrients
   b) Water quality and supply
   c) Feed toxins
   d) Feed contaminants
   e) Change of diet
   f) Drugs-antimicrobials, growth promoters
7. Biological
   a) Insects
   b) Rodents
   c) Wild birds
   d) Opportunistic infectious agents
8. Social
   a) Population density
   b) Sex ratio.

Disease Transmission

Infectious diseases do not occur by chance. It is required for a harmful infection environment.

The disease causing organisms are introduced or transmitted to a poultry flock in a variety of ways.

- Carriers within a flock
- Birds recently acquired from an out side flock
- Eggs from infected breeder flocks
- Human feet, hands and clothes
- Dust, feathers and manure on equipment and supplies, such as trucks, coops and egg filler flats
- Pet birds, predators, rodents, flies and insects
- Live or contaminated vaccines
- Contaminated feed/feed ingredients
- Water
- Air.

Disease Prevention and Control

Disease control involves all the measures designed to prevent or reduce the incidence, prevalence and consequences of the disease. The basic approach in controlling disease is to identify the weak points of diseases and break the weakest links in the chain of transmission (Pattison, 1993). Broadly these measures include: 1) controlling the reservoir or source of infection, 2) breaking the routes.

Controlling the Reservoirs and Source of Infection

If the first link in the chain of causation (i.e. the disease agent) is deemed to be the weakest fink, logically, the most desirable control measure would be to eliminate the reservoir source. This includes all the measures directed to reduce the quantity of the agent available for dissemination.

1. Early identification of reservoir/source of infection
2. Treatment of affected birds
3. Epidemiological investigation
4. Isolation (separation) of infected birds from healthy birds
5. Serological testing and removal of carriers
6. Hygienic disposal of dead birds, litter and hatchery waste

7. through cleaning, washing and disinfection of farm houses, equipment
8. Feed and Water quality control
9. Hygienic hatching of chicks.

Breaking Transmission Routes

A major aspect of communicable disease control relates to "breaking the chain of transmission". Microbes travel from place to place by chicken crates, egg filler flats, trucks, other equipment and people. They can be transmitted from one flock to the other through workers. The organisms are found on the skin and in the digestive and respiratory organs of dogs, cats, rats, mice and wild free flying birds, Flies, beetles and other insects are also notorious spreaders of disease causing organisms.

Blocking the routes of transmission imply on control of environmental factors, that is, to bring about an adjusted equilibrium between host and environment through encouraging some ecological influences and inhibiting others (Woodger.1996).

- Human traffic control, such as locking the doors and banning the visitors.
- Providing foot dips (at the entrance of all houses), wheel dips at the entrance of the farm
- Aerial and water sanitation
- Pest control (flies, mosquitoes, ants, rodents, litter beetles, lice, bed bugs, wild birds, cockroaches, spiders, mites and weeds)
- All-in-all out farm management
- Restriction of movement of personnel, feed, vehicles and farm equipment
- Personnel hygiene
- Avoiding of on farm trading.

7

Effect of Dietary Protein Concentrates

Necrotic enteritis (NE) is a widespread and economically important bacterial disease in modern broiler flocks (Van Der Sluis, 2000). The subclinical form of the disease is more common than clinical outbreaks in broiler flocks (Kaldhusdal, 2000). The condition is not usually detected due to the absence of clear clinical signs; therefore, it is not treated and prevails unnoticed apart from a poor growth performance (Lovland and Kaldhusdal, 2001), wet litter conditions (Williams, 2005), and the possible contamination of poultry products for human consumption (Craven et al., 2001). The financial cost of NE has been estimated to be US $2.6 billion per year to the world's poultry industry (Van Der Sluis, 2000).

In subclinical NE, the major pathological changes occur in the small intestine (Gholamiandehkordi et al., 2007) and the liver. Intestinal *Clostridium perfringens* counts (Kaldhusdal and Hofshagen, 1992) and intestinal *C. perfringens* α-toxin levels are also increased. The causative organism, *C. perfringens*, is ubiquitous and found in soil, dust, feces, feed, used poultry litter, the intestines of most healthy animals, and humans. *Clostridium perfringens* also is a commensal bacterium of chicken intestines (Ewing and Cole, 1994) and so there are further predisposing factors that alter the intestinal balance in favour of the proliferation of the causative bacteria and allow them to migrate to the upper intestines. Apajalahti et al. (2001) identified the diet as the strongest determinant of the cecal bacterial community, so the diet composition may influence the susceptibility of broiler chickens to subclinical NE, although there is a lack of direct experimental evidence. However, there is evidence that different dietary protein sources affect the proliferation of *C. perfringens* within the cecum (Drew et al., 2004) and in the ileum when birds are orally dosed with these bacteria. Also, studies have indicated that the dietary amino acid balance may influence the proliferation of *C. perfringens.*

Most of the published experimental information on NE has reproduced the disease by dosing the birds with pathogenic strains of *C. perfringens* isolated from clinically diseased birds. Although the various protocols differ, most of these experiments involved keeping a small number of birds in cages with frequent dosing of the pathogen. The majority of the published work that has examined dietary factors that predispose birds to NE has also used much higher inclusion rates of the individual feedstuffs than would be used in proprietary feeds. All of these conditions are different from practical broiler chicken rearing methods.

However, Lovland et al. (2003) reported that a mild or subclinical NE initiates spontaneously without dosing the birds with pathogenic *C. perfringens.* This innate infection usually occurs when the appropriate predisposing factors (unmedicated diets, putative predisposing feeding regimens, housing birds on litter) are provided. This method of reproducing subclinical NE therefore provides the possibility of studying the effects of dietary factors under conditions that are directly related to commercial production methods.

There is a need to determine whether dietary protein supply affects the incidence of subclinical NE in practical broiler growth conditions. Dietary protein supply is a major variable in the formulation of poultry feeds and local and world price variations can result in large changes in the sources and quality of the protein used. Soy is the most frequently used protein concentrate in poultry feed formulations. Experimental evidence suggests that high contents of dietary fish meal increase the intestinal *C. perfringens* populations in pathogen-dosed birds (Drew et al., 2004). Wilkie et al. (2005) identified that birds fed diets with high potato protein inclusion rates also had higher counts of *C. perfringens* when compared with birds fed other plant protein diets. However, the effect of potato protein and fish meal, in comparison to soybean meal, on naturally occurring spontaneous subclinical NE has not been investigated. Therefore, the objective of this experiment was to identify the effects of 3 different dietary protein concentrates: potato protein concentrate (PPC), fish meal, or a mixture of soybean meal and full-fat soy in nutritionally complete diets, with similar protein contents (CP 21%), on the incidence of subclinical NE in male broiler chickens reared in comparatively large flocks and in conditions close to practical production methods.

Materials and Methods

Dietary Treatments and Experimental Design

Three practical broiler grower diets were formulated to be nutritionally complete for macronutrients (Aviagen, 2007a) and the vitamins and minerals were provided with a proprietary vitamin and mineral premix that either met or exceeded the NRC (1994) recommendations for broiler chickens between 15 to 31 d of age. Diets were corn-based, but a major proportion of the additional protein supply was provided by 1 of 3 protein concentrates: PPC (CP 79%) provided 58% of the total CP in the potato-based diet and fish meal (South American origin; CP 66%) provided 58% of CP in the fish-based diet.

The soy protein-based diet was composed of a mix of soybean meal (dehulled; CP 48%) and toasted full-fat soy (CP 36%) and these 2 soy proteins together provided 61% of the total CP in the diet. All 3 diets were formulated to have similar contents of calculated ME (3.1 mcal/kg), CP (21%), Lys (1.3%), Met and Cys (0.94%), and Thr (1.02%). However the chemical analysis at the end of feed processing (mixing and pelleting) indicated that there were differences in amino acid, starch, lipid concentrations, and trypsin inhibitor activity between the treatment diets. No antibiotic growth promoters or anticoccidial drugs were used in the diets. The diets were all provided as 3-mm-diameter pellets. Treatment diets were compared using 6 replicate pens for each treatment. Eighteen pens were used within 2 adjacent environmentally controlled rooms. The dietary treatments were randomly allocated to pens within 6 positional blocks (3 per room).

Broiler Chicken Management and Feeding

The broiler chicken experiment was conducted under the guidance of the Research Ethics Committee of Harper Adams University College. A total of 1,300 one-day-old male Ross 308 broiler chickens were reared in a solid-floored pen as a single flock in an environmentally controlled house. Adequate feeders and drinkers were provided for the age and the number of birds. For the first 15 d, birds were fed a proprietary, nutritionally complete broiler starter feed formulation that was unmedicated (without in-feed antibiotics or coccidiostat). On 15 d post-hatch, the birds were weighed and 70 birds were randomly allocated to each of the 18 pens (altogether 1,260 birds were allocated to the experiment). The floor area of (1.5 m × 3 m) each pen was covered with wood shavings that comprised 4 parts new wood shavings to 1 part reused litter material from a previous poultry flock that did not have a history of clinical NE but some subclinical NE and subclinical coccidiosis would have been expected. The birds were fed the experimental diets for the following 16 d. Feed and water were provided ad libitum during the experimental period. Feed levels were kept low in the hanging tube feeders to minimise wastage. Light was provided for 23 h per day with controlled temperature and humidity. The growth and feed intakes of the birds were recorded over the experimental period. Mortality was recorded daily.

Data Collection

Lesion Scoring

On 27, 28, 29, and 30 d of age, 8 birds from each replicate pen (2 birds per day) were selected at random and killed by cervical dislocation and 5 mL of blood was collected in glass tubes from the jugular vein. The liver of each bird was examined for the presence or absence of lesions of hepatitis or cholangiohepatitis that were consistent with the pathological changes of NE (Lovland and Kaldhusdal, 1999; Sasaki et al., 2000). The intestinal tract was removed and the duodenal loop was separated and retained. Three further 8-cm sections were taken from the rest of the small intestine; the first of these sections was the proximal jejunum, the second section was either side of Meckel's diverticulum (mid small intestine), and the third section was the terminal ileum.

All 4 sections (including the duodenal loop) were immediately incised, washed in normal saline, and the mucosal surfaces were inspected and necrotic and hemorrhagic lesions were scored. Proximal jejunal sections were directly stored at "20°C. Sections (1 cm^2) from the jejunum that had focal necroses were fixed in 10% formal saline and after 2 d, these sections were rinsed with water and placed in 70% ethanol for histopathological diagnosis. Duodenal lesions were scraped, smeared on glass slides, and 16 representative samples from each diet (4 samples per day) were directly plated (in duplicate) onto blood agar and incubated anaerobically at 37°C for 48 h.

A 5-point scoring system used by Gholamiandehkordi et al. (2007) was planned to score intestinal lesions, but in the present experiment, all observed lesions were focal necroses with a size of 1- to 10-mm diameter and the rest of the intestinal sections had no necrotic lesions. Therefore, only 2 scores were used: 0, no lesions and 1, focal necroses (1- to 10-mm diameter). Focal lesions varied from 1- to 2-mm white foci to around 10-mm-diameter mucosal depressions with yellowish green mucoid material laid in the periphery. Whenever such lesions appeared alongside the hemorrhagic lesions, these intestinal sections were recorded as necrotic lesions-positive regardless of the level of the other lesions.

In addition, all of the intestinal sections were scored for hemorrhagic lesions regardless of the incidence of necrotic lesions. Sections with no hemorrhagic lesions were scored as zero, petechial hemorrhages (pinpoint hemorrhagic foci of 1- to 2-mm diameter) of the intestinal mucosa were given score 1, purpura hemorrhages (hemorrhagic foci of 3-mm diameter) as score 2, ecchymotic hemorrhages (blotchy or irregular hemorrhages up to 1 to 2 cm in size) as score 3, coalesced paint brush hemorrhages as score 4, and extravasation as score 5.

Serum Antibodies for α Toxin

The collected blood was allowed to clot for 8 to 10 h at room temperature. The serum was then separated from the coagulum and stored at –20°C. Eight serum samples from each pen were tested for antibody levels developed against the α toxin of *C. perfringens* using an indirect ELISA (Heier et al., 2001; Lovland et al., 2003). Each well of the Nunc immunoplates (F96 Maxisorp, 735-0083, Thermo Fisher Scientific, Rochester, NY) was filled with 100 μL of known antigen, phospholipase C type XIV (Sigma-Aldrich P 4039, St. Louis, MO), 10 μg/mL in sodium carbonate and bicarbonate coating buffer at pH 9.6. The plates were incubated at 4°C overnight to coat the wells with the antigen. The plates were then washed 3 times with PBS at pH 7.2, with 0.05% Tween 20 (PBST). A volume of 150 μL of 1% BSA in PBST was added to each well as a blocking buffer and incubated at 37°C for 2 h. The plates were then washed again in PBST and left for 1 h at room temperature to dry. The antigen-coated plates were individually sealed, packed, and stored at 4°C.

Serum samples were diluted at 1:250 in PBST and 50 μL of each sample was added into duplicate wells. The plates were left at 4°C overnight and then washed 3 times with PBST. Rabbit anti-chicken immunoglobulin IgY whole molecule conjugated with alkaline phosphatase (Sigma-Aldrich A9171) was diluted (10^{-4}) in blocking buffer and 50 μL of diluted anti-chicken antibodies was added to each well and then the plates were incubated at 37°C. After 2 h, the plates were washed with PBST.

The plates were incubated with 150 μL of para-nitrophenyl phosphate (Sigma-Aldrich P 7998) for 1 h at 37°C. The reaction was stopped with 50 μL of 2 *M* NaOH. The optical density (OD) was read at 405 nm using a microplate reader (Bench Mark 170-6850, Bio-Rad, Hercules, CA). Resulting OD values were pooled for each pen.

Enumeration of C. Perfringens

Clostridium perfringens that were colonised on the mucosal surface of the proximal jejunum were quantified using the BIO K 086-*C. perfringens* antigen detection kit (Bio-X Diagnostics, Jemelle, Belgium) in a double antibody sandwich ELISA (McCourt et al., 2005). Four birds from each replicate pen were used to quantify *C. perfringens* in the mucosa of the jejunum. Gut samples were thawed at room temperature and washed 3 times in PBS. Then the luminal surface of the proximal jejunum was scraped with a sterile surgical blade and half of the scraping was weighed into a microcentrifuge tube and diluted (2×) with dilution buffer (Bio-X Diagnostics) and mixed vigorously before being allowed to settle for 10 min. (The other half of the scraping was retained for coccidia oocyst counts.) Each sample was filled into 2 adjacent wells (one coated with specific monoclonal antibodies for *C. perfringens* and the other coated with non-specific antibodies) of the ELISA plates. The plates were incubated at 21 ± 5°C for 1 h and then washed 3 times with washing solution (Bio-X Diagnostics). Subsequently, peroxidase-labeled anti-*C. perfringens*-specific monoclonal antibodies were added to each well and then incubated at 21 ± 5°C for 1 h and washed 3 times. Freshly mixed substrate (hydrogen peroxide) and chromogen (tetramethyl benzidine) was then added to each well. After 10 min, 1 *M* phosphoric acid was added to each well and the resulting OD was read at 450 nm using a microplate reader (Bench Mark 170-6850, Bio-Rad). A broth culture of *C perfringens* was used for serial dilutions for plate counts and at the same time for fixation in the dilution buffer (Bio-X Diagnostics) for ELISA (samples for ELISA were immediately frozen). After the plate counts (each dilution was counted 4 times), the ELISA was performed and a standard curve ($r = 0.945$) was created (Bench Mark 170-6850). This was used to convert the OD values into bacterial counts and the values were pooled for each pen.

Enumeration of Coccidia Oocysts

Eight birds from each replicate pen were sampled for coccidia oocyst counts in the proximal jejunum. Scrapings as taken for the *C. perfringens* counts were diluted (2×) with saturated NaCl and then placed into both sides of a McMaster counting chamber (Chalex Corporation, Wallowa, OR). The number of coccidia oocysts (all species found in the sample) was counted at 10 × 10 magnification. The counts were pooled for each pen before statistical comparison.

Feed Analysis

Dry matter content of the experimental feeds was determined by drying the samples to constant weight at 105°C in an oven. Ash content was determined by AOAC method 942.05 (AOAC International, 2000). Nitrogen was determined with an automatic analyser (Leco FP-528 nitrogen, Leco Corp., St Joseph, MI) by AOAC 968.06 (Dumas method) using EDTA as the standard and the protein content was calculated as nitrogen × 6.25. Gross energy in the feed was determined with an adiabatic bomb calorimeter (model 1261 isoperibol, Parr Instrument Co., Moline, IL) using analytical grade sucrose as the standard. The fat content was determined with the AOAC 920.39 method using a Soxtec 1043 extraction unit. Soluble and insoluble non-starch polysaccharides (NSP) were determined using AOAC methods 991.43 and 985.29 (for soluble and insoluble polysaccharides, resistant starch, and lignin), AACC methods 32-07 and 32-05 (for total dietary fibre), and AACC method 32-21(for soluble dietary fibre; Megazyme K-TDFR assay kit, Megazyme International Ireland Limited, Wicklow). Total starch contents of the diets were determined with AOAC method 996.11 for high amylose cereal starch and AACC method 76.13 for gelatinised starch (Megazyme K-TSTA assay kit). Trypsin inhibitor activity was determined by measuring the inhibition of milligrams of bovine trypsin for a gram of sample according to the Kakade method as modified by Smith et al. (1980). Crude protein and gross energy were determined in triplicate samples and all the other analyses were performed in duplicate.

Dietary amino acid samples were oxidised with a hydrogen peroxide-formic acid-phenol mixture. Excess oxidation reagent was decomposed with sodium metabisulphite. The oxidised sample was hydrolysed with 6 *M* hydrochloric acid for 24 h. The hydrolysate was adjusted to pH 2.20, centrifuged, and filtered. The amino acids were separated by ion exchange chromatography (Biochrom 20 analyser, Amersham Pharmacia Biotech, Pittsburgh, PA) and determined by reaction with ninhydrin using photometric detection at 570 nm (440 nm for proline). For tryptophan analysis, the samples were hydrolysed with 4.2 *M* sodium hydroxide for 23 h. The hydrolysate was adjusted to pH 2.20, centrifuged, and filtered. The tryptophan contained in the hydrolysate was then detected by ion exchange chromatography (Biochrom 20 analyser, Amersham Pharmacia Biotech) and determined by reaction with ninhydrin using photometric detection at 570 nm.

Statistical Analysis

The effects of the protein sources on growth performance, serum antibody levels for α toxin, and the number of coccidia oocysts and

C. perfringens in the proximal jejunum were compared using a randomised block ANOVA (GenStat Release 10.1, Lawes Agricultural Trust, Rothamsted Experimental Station, Harpenden, UK) using the pen as the experimental unit. Individual treatment differences were compared by a protected least significant difference test using a probability of less than 0.05. The data obtained for the incidence of intestinal necroses, intestinal hemorrhages, liver lesions, percentage of birds that had mucosal *C. perfringens*, and the pen mortality rates were compared using a non-parametric $\div^2$ test.

Results

The proximate nutrient compositions of the experimental diets were approximately similar. However, the potato diet had higher trypsin inhibitor activity and lower crude fat content than that of fish or soy product-containing diets. Trypsin inhibitor activities of potato-, soy-, and fish-based diets were 3.88, 1.51, and 0.82 mg/g, respectively. The soy diet had the highest starch content among 3 diets. The fish diet had a lower soluble NSP content compared with the potato or soy diets. The insoluble NSP content was highest in the potato diet. The potato diet also had higher levels of valine, leucine, tyrosine, and phenylalanine, whereas the fish diet had high levels of glycine, lysine, and methionine. The soy diet had high Glu and Asp levels compared with the potato and fish diets.

The weight gains and feed intakes of the birds fed the potato or fish diets were significantly lower than those fed the soy diets. Weight gain:feed intake ratio and mortality rate were not affected ($P > 0.05$) by dietary treatment. The birds fed the potato diet had a higher ($P < 0.01$) incidence of necrotic lesions in the duodenum and proximal jejunum compared with the birds fed the soy diet. The necrotic lesion incidence in the birds fed the fish diet was intermediate. No necrotic lesions were identified in the mid or distal sections of the small intestine, in any of the treatment groups, although pseudo-membrane formation was observed in a large proportion of birds (data not presented). There was a higher ($P < 0.05$) level of serum antibodies for *C. perfringens* α toxin in birds fed the potato diet compared with the other 2 diets.

Double hemolytic colonies on blood agar were confirmed as *C. perfringens* with biochemical tests and all gram-stained smears predominantly had relatively large (4 to 6 μm long) gram-positive rods with blunt ends existing as singles or up to 4 loosely attached chains or clusters. They were morphologically similar to *C. perfringens* isolated on blood agar plates. Similar bacteria were also attached to damaged villi tips in histological examination of jejunal sections. Fifty-six percent of the birds had *C. perfringens* colonised on the mucosa of the proximal jejunum, but there were no ($÷^{2}$2 df = 2.14, P = 0.343) treatment differences (66.7, 45.8, and 54.2% for the potato, fish, and soy diets, respectively) in the percentage of birds that had *C. perfringens* colonised on the jejunum or in their counts.

The birds fed potato or fish diets had a higher ($P < 0.001$) proportion of moderate to severe (scores of 3 to 5) duodenal hemorrhages (64 and 77%, respectively) than the birds fed the soy diet (53%). There were no treatment differences in hemorrhagic lesions in the proximal jejunum and the mid small intestine even though there was a relatively high overall incidence. The birds fed the potato diet had a higher proportion ($P < 0.001$) of hemorrhagic lesions than the soy- or fish-fed birds in the distal ileum of the small intestine. The birds fed the potato diets also had a higher ($P < 0.001$) incidence of liver lesions (96%) than those fed the fish (54%) or soy (46%) diets. Mortality tended ($P < 0.1$) to be higher in potato-fed birds. Although none of the birds had clinical coccidiosis, the birds fed the soy diet had a higher ($P < 0.01$) number of coccidia oocysts than those fed the fish diet.

Discussion

The growth performance of the birds in the experiment was comparable with commercial levels (Aviagen, 2007b). A proportion of the birds in all 3 dietary treatment groups had some evidence of the subclinical form of NE. This confirms the finding of Lovland et al. (2003) that a spontaneous *C. perfringens* infection, with some predisposing factors, is sufficient to reproduce subclinical NE in a proportion of birds in relatively large floor-reared flocks. At the beginning of the experimental period, changes in the environment (used litter mixed into fresh litter might have added some coccidia oocysts and *C. perfringens*) with simultaneous dietary changes would have predisposed the birds to subclinical NE, as described by Parish (1961). The mortality percentages found in the experiment were within the ranges reported in subclinical disease and were within the range commonly found in commercial broiler flocks at this age.

The birds fed the potato protein diet, in comparison to the soy diet, had a significantly greater incidence of necrotic lesions in the duodenum and the jejunum. The pseudo-membrane formation (mucosa covered with golden brown to greenish yellow loosely adherent material) that was observed in the distal and middle small intestine suggests that these birds might have had necrotic lesions in the distal and middle small intestine before sampling and they were recovering from these lesions. In addition to intestinal lesions, the birds fed the potato protein diet had a significantly greater incidence of liver lesions, higher serum antibody levels for *C. perfringens* α toxin, and a tendency for a relatively small increase in the flock mortality rate. The proportion of birds that had *C. perfringens* colonised on the jejunal mucosa tended to be higher in the potato diet treatment. There was no evidence of an increased incidence of *C. perfringens* organisms colonised on the gut wall, confirming the findings of McReynolds et al. (2004),who identified no relationship between the incidence of necrotic lesions and the total *C. perfringens* population of the jejunum.

Higher coccidia oocyst counts were found in the soy-fed birds compared with those fed the fish diet. This indicates that the subclinical coccidiosis did not directly contribute to the pathological lesions in the intestine. However, coccidiosis is a factor that predisposes birds to NE and has been identified to have a synergistic relationship with *C. perfringens* during the development of experimental NE. Coccidia multiplication stages in the intestinal epithelium initiate gut mucosal damage but may then be excluded from their proliferation sites when *C. perfringens* colonise the damaged intestinal epithelium further destructing the enterocytes (Williams et al., 2003). Also, the diphtheritic membrane formation could impede the intraluminal dissemination of extracellular coccidial stages.

The 3 diets used in the present experiment had a similar proximate nutrient composition, despite several nutritional differences. Amino acid imbalance has been suggested as a risk factor for subclinical NE. Although there is evidence of an effect of dietary glycine on *C. perfringens* counts in the small intestine and cecum when the birds are orally dosed with the bacterium, in this experiment, there were no major differences in the glycine levels in potato (9.4 g/kg) and soy (9.0 g/kg) diets. Bacterial degradation of aromatic amino acids, such as tyrosine, produces phenolic and aromatic compounds that are toxic to the intestinal epithelial cells; therefore, the higher levels of aromatic amino acids in the potato diet than in the other 2 diets could have contributed to the initial damage of the epithelial cells.

The growth of *C. perfringens* under in vitro conditions needs 11 amino acids. Muhammed et al. (1975) suggested that some amino acids, such as methionine, stimulate the growth of the bacterium, but Dahiya et al. (2007b) found a reduction of *C. perfringens* populations in the cecum and the ileum of the birds fed high Met-containing diets.

However, in the present experiment, there was no difference in Met levels between the soy and potato diets (4.9 g/kg), although the fish diet had a slightly higher level of Met (6.2 g/kg). The differences in amino acid balance between the diets do not appear to be correlated to the observed differences in subclinical NE.

Amino acid availability was not determined in the present experiment, but may have differed between the 3 treatment groups. The determined trypsin inhibitor activity was higher in the PPC-based diet than in the soy- and fish-based diet. A high trypsin inhibitor activity in PPC has been found in other studies (Lee et al., 1985). Therefore, it is possible that the PPC-based diet may have had a low protein digestibility and increased the protein entering the distal small intestine and ceca of the chickens when compared with the soy and fish diets. Taciak and Pastuszewska (2007) found that a PPC-based diet with low protein digestibility modified the cecal fermentation of rats, giving higher levels of ammonia and butyrate instead of acetate and propionate in comparison to a soybean meal diet. High cecal butyric acid concentrations have also been observed in birds with subclinical NE.

The protease inhibitors in PPC can vary considerably depending on variety and quality of potatoes used (Jadhav and Kadam, 1998) and also with the method of processing. Heat coagulation at low pH is the method mainly used for industrial purposes and this method of processing can denature the proteins considerably, resulting in a very low nitrogen solubility at low pH (Knorr, 1982). However, the protease inhibitors are less heat-labile, more soluble (at the whole pH range in the intestine), and have lower molecular weights (5 to 25 kDa) when compared with the storage proteins of potato. Potato also contains high molecular weight (85 kDa) protease inhibitors, which can be cleaved into several functional, lower molecular weight fragments by intestinal trypsin (Walsh and Strickland, 1993). The high trypsin inhibitor activity in the PPC-based diet in the present experiment indicates that some protease inhibitor activity remained in PPC.

The dietary protease inhibitors could also have directly influenced the incidence of NE by intensifying the activity of *C. perfringens* toxins. Trypsin inhibitors stabilise functional properties of phospholipase c (α toxin) for longer when they are directly bound with the toxin (Sof'ina and Rakhimov, 1998), which could facilitate the destruction of phospholipid cell membranes. In addition, the high trypsin inhibitor content and some other possible protease inhibitors in the PPC-based diet could have inhibited pancreatic trypsin, which is able to cleave α (phospholipase c) and â 2 (NetB) (Fisher, 2006) toxins to inactivate the toxic activities. These are 2 major toxins in *C. perfringens* toxin type A, which has been mostly implicated in the etiology of NE (Engstrom et al., 2003) Therefore, the higher trypsin inhibitor activity could have aggravated toxin-induced intestinal necroses in the birds fed PPC-based diets than in the birds fed the fish and soy diets.

The glycoalkaloid content of industrial PPC has been estimated to be 257 mg/kg and this exceeds the widely accepted safety limit of 200 mg/kg in humans. Glycoalkaloids are not heat-labile and also have the ability to disrupt cell membranes, which could damage the intestinal epithelium. *Clostridium perfringens* usually colonise on damaged intestinal epithelium (Parish, 1961). Therefore, the glycoalkaloids together with subclinical coccidiosis could have predisposed the *C. perfringens* damage, which subsequently exacerbated leading to necroses of the intestinal mucosa. Dietary potato glycoalkaloids are known to damage the cells in contact tissues such as liver and blood (Smith et al., 1996); therefore, when they accumulate in the liver, glycoalkaloids could have contributed to the high incidence of liver lesions observed in potato diet-fed birds when compared with the birds fed the other 2 diets.

The potato protein diet had a higher level of insoluble NSP but a low soluble NSP content compared with the soy diet. The corn content in the potato diet was higher than in the soy diet. Corn has a considerable content of insoluble NSP (70 g/kg) in the total NSP (76.3 g/kg; Meng and Slominski, 2005), which is probably the reason for the higher NSP level in the PPC-based diet. The published literature is contradictory on the effect of NSP.

The inclusion of cereals with high levels of NSP such as wheat (Branton et al., 1987), barley (Kaldhusdal and Hofshagen, 1992), and rye (Craven, 2000) increases the incidence of NE. Carbohydrase enzyme addition mitigates the negative effects of a *C. perfringens* challenge (Jia et al., 2009). Conversely, Riddell and Kong (1992) could not demonstrate these effects and Branton et al. (1997) concluded that added complex carbohydrates and fibre in broiler rations reduced the intestinal lesions of NE. However, it is unlikely that the relatively small differences in NSP contributed to the incidence of subclinical NE in this experiment.

The oil content of PPC is very low (0.52 g/kg) and this resulted in a low oil content in the potato-based diet. The low oil content in the potato diet could have been a contributory factor for the increased incidence of subclinical NE. High lipid intakes have been associated with an increased bile acid synthesis and excretion (Reddy et al., 1977). The antibacterial effect of bile salts has been demonstrated by Inagaki et al. (2006). However, Knarreborg et al. (2002a) identified *C. perfringens* and *Enterococcus faecium* as the 2 bacterial species that had the highest bile acid hydrolase activity in the small intestinal flora of broiler chickens. Dietary soy oil has been shown to significantly reduce the population of *C. perfringens* in the ileal microflora of broiler chickens compared with dietary animal fat (Knarreborg et al., 2002b). In this experiment, the oil composition of fish meal would have differed from other 2 diets, but the soy and potato diets had similar sources (corn and soy) of dietary oils. Therefore, the oil composition could not have contributed much to the disease incidence.

In conclusion, this experiment successfully reproduced spontaneously occurring subclinical NE in broiler chickens fed all 3 experimental diets. There was a significant increase in the incidence of subclinical NE in the birds fed potato protein within a nutritionally complete diet in comparison to soy- or fish-based diets. The differences in NE incidence were not consistent with the relatively small differences in amino acid balance or NSP content, but the high trypsin inhibitor activity, low oil content, and possible heat-resistant toxic compounds of the potato protein diet could have contributed to the increased incidence of subclinical NE in the PPC-based diet-fed birds.

Campylobacter Infection

Campylobacter spp. are bacteria that commonly infect a broad range of livestock species, pets and wild animals. In poultry they tend to multiply in large numbers in the hindgut, principally in the caecae. Campylobacters are a significant cause of enteritis in man. Infected poultry are a potential reservoir of this zoonosis.

Campylobacter jejuni is the commonest species found in poultry. All campylobacters are delicate organisms that survive for relatively short periods outside the host unless protected by organic material, biofilm or engulfed by protozoa.

Campylobacter jejuni infection is not currently considered to be pathogenic in poultry though a *Campylobacter*-like organism is considered to be the cause of 'Vibrionic Hepatitis'.

There are indications that plantar pododermatitis, carcase quality and litter quality are better on farms which tend to have *Campylobacter*-

negative stock. The reason for this is unclear. It may be that management that favours dry litter reduces the risk of infection and/or transmission within the flock. There is an annual cycle with increased risk of infection in the summer months in some countries.

Signs:

- None.

Post-mortem lesions:

- None.

Diagnosis

- Isolation of the organism from caecal contents, cloacal swabs or composite faeces. The organism is sensitive to air so swabs should be collected into transport medium and other samples placed in airtight containers with minimal airspace. Samples should be tested as quickly as possible after collection.

Treatment

- Not required on clinical grounds.

Prevention

- In principle, housed poultry can be maintained free of *Campylobacter* infection by consistent application of excellent biosecurity. Key aspects of this include effective sanitation of drinking water, sourcing of water from high quality supplies, avoidance of contact with pets and other farmed species, good hand hygiene by stockmen, and changing of overalls and boots on entering bird areas.

In practice the success of this will also depend upon the degree of environmental contamination by the organism. For this reason it may be difficult to stop the spread of infection between houses once it becomes established in one house. Many infections are introduced during thinning or other forms of partial depopulation. Insects and rodents may act as a means of transfer of the infection from the general environment into the poultry buildings. Research is ongoing on the development of vaccines, phage treatments and competitive exclusion approaches, as well as processing plant technologies to reduce carcase contamination.

Candidiasis, Moniliasis, Thrush

A disease of the alimentary tract of chickens, turkeys, and sometimes other birds and mammals, characterised by thickening and white plaques on the mucosa, especially in the crop but sometimes in the proventriculus, intestine and cloaca, and associated with gizzard erosion. The cause is a fungal yeast, *Candida albicans* and the condition is seen worldwide. Morbidity and mortality are usually low. The route of infection is normally oral and the organism is often present in healthy animals with disease occurring secondary to stress and poor hygiene. The fungus is resistant to many disinfectants.

Signs:

- Dejection.
- Poor appetite.
- Slow growth.
- Diarrhoea, possibly confused or masked by signs of the primary disease.

Post-mortem lesions:

- White plaques in mouth, oesophagus, crop, occasionally proventriculus and intestine.
- Raised focal lesions may slough into lumen as caseous material.

Diagnosis

- Lesions, histopathology, microscopic examination of a digested smear (heat in 10% potassium hydroxide) to demonstrate the hyphal forms of the yeast in the tissues. Colonies of this fungus appear as white to ivory colour, smooth and with a yeasty smell.

Treatment

- Nystatin (100 ppm in feed) for 7-10 days, copper sulphate (1 kg/tonne feed) for 5 days, or copper sulphate 1gm/2 litre water for 3 days if approved locally.

Prevention

Avoid excessive use of antibiotics and other stressors. Ensure good hygiene, proprionic acid, sodium or calcium proprionate at 1 kg per tonne continually. A finely divided powder of copper sulphate (where approved) at 200gm/tonne continually or to 14-16 weeks in replacement pullets.

Control of *Candida* through drinking water is sometimes practised with chlorination (e.g. Chlorox, sodium hypochlorite) at 5 ppm. This is economical and effective. It should be repeated periodically. Take care to provide fresh clean feed and water, uncontaminated by fungi.

Cannibalism, Feather Pecking

A complex multifactorial behavioural problem of poultry and game birds seen worldwide. Morbidity is usually low but mortality is high among affected birds. Predisposing factors include overcrowding, excessive light intensity or variation (e.g. through shafts of light in the house), high temperatures, nutritional deficiencies, feed form (mash takes longer to consume than pellets), tenosynovitis and other diseases affecting mobility, boredom, and strain of bird.

Signs:

- Pecking at feet (especially young chicks) and vents (adult layers and turkey poults 8-12 days old), head, face, wings.
- Feather-pulling.

Post-mortem lesions:

- Skin wounding related to particular signs exhibited.
- Generalised anaemia.

Diagnosis

- Age, distribution of lesions, anaemia. Differentiate from bacterial dermatitis, post-mortem cannibalism.

Treatment

- Correct any husbandry problems. Soluble multivitamins and/ or methionine may be of some benefit in some circumstances. Beak trimming may be necessary. If so it should be carried out carefully by trained operators, complying with local regulations and any relevant codes of practice.

Prevention

- Proper density and temperature, low light level, control ectoparasites. Provision of a diet that closely matches the nutritional requirements of the stock concerned.

Cellulitis

Cellulitis is literally an inflammation of connective tissues. It typically occurs between skin and muscles and between muscles and may be an incidental finding in a range of conditions. However its main importance is as a cause of condemnation in meat poultry, particularly broiler chickens. In the USA it is called 'Inflammatory Process'. The condition is caused by infection of, often minor, skin wounds by particular strains of *E. coli*, which can replicate in the tissues.

Signs:

- Affected flocks tend to have poorer than average productivity and uniformity, but the affected birds are not readily detectable prior to slaughter.

Post-mortem lesions:

- Typically it presents as exudate ranging from liquid and pale cream pus to yellowish solid plaques of caseous material under the skin of the abdomen and/or in the leg. Many affected birds have no other lesions and are reasonably well grown. Many meat inspectors become skilled at detecting subtle differences in skin colour in the affected birds.

Diagnosis

- Typical lesions.

Treatment

- Treatment would not be possible if the problem is identified at a final depletion. If identified at a thinning there may be time for antibacterial treatment to have some benefit for those birds in the early stages of the problem.

Prevention

- Toe scrapes at 15-25 days of age when feather cover is poor are the most likely predisposing factors. Careful flock management with a view to reducing toe wounds has the greatest impact in controlling cellulitis. Routine monitoring of skin damage at about 25 days of age may be helpful in fostering good practices, though most of the birds showing toe scrapes will not go on to develop cellulitis.

Chicken Anaemia

A viral disease of chickens caused by Chicken Anaemia Virus or CAV. Prior to confirmation that it is in fact a virus it was known as Chicken Anaemia Agent or CAA. Mortality is typically 5-10% but may be up to 60% if there are predisposing factors present such as intercurrent disease (Aspergillosis, Gumboro, Inclusion body heptatitis, etc.) or poor management (e.g. poor litter quality).

Transmission is usually vertical during sero-conversion of a flock in lay, lateral transmission may result in poor productivity in broilers.

The virus is resistant to pH 2, ether, chloroform, heat (70°C for 1 hour, 80°C for 5 minutes) and many disinfectants even for 2 hours at 37°C. Hypochlorite appears most effective *in vitro*.

Signs:

- Poor growth.
- Pale birds.
- Sudden rise in mortality (usually at 13-16 days of age).
- No clinical signs or effect on egg production or fertility in parent flock during sero-conversion.

Post-mortem lesions:

- Pale bone marrow.
- PCV of 5-15% (normal 27-36%).
- Atrophy of thymus and bursa.
- Discoloured liver and kidney.
- Gangrenous dermatitis on feet, legs wings or neck.
- Acute mycotic pneumonia.

Diagnosis

- Gross lesions, demonstration of ongoing sero-conversion in parent flock, virus may be isolated in lymphoblastoid cell line (MDCC-MSB1).

Treatment

- Good hygiene and management, and control of other diseases as appropriate, may be beneficial. If gangrenous dermatitis is a problem then periodic medication may be required.

Prevention

Live vaccines are available for parents, their degree of attenuation is variable. They should be used at least 6 weeks prior to collecting eggs for incubation. Their use may be restricted to those flocks that have not sero-converted by, say, 15 weeks.

Immunity: there is a good response to field challenge (in birds over 4 weeks of age) and to attenuated live vaccines. Serology: antibodies develop 3-6 weeks after infection, and may be detected by SN, Elisa, or IFA.

Chlamydiosis, Psittacosis, Ornithosis

An infection of turkeys, ducks, psittacines, pigeons, man, rarely chickens, caused by *Chlamydia psittaci*, a bacterium of highly variable pathogenicity. It is a 'Scheduled Disease' rarely diagnosed in UK, but occurring probably worldwide. Morbidity is 50-80%, mortality 5-40%. It is transmitted by contact, faecal dust and wild bird carriers, especially pigeons and robins. Egg transmission does not occur. Elementary bodies are highly resistant and can survive in dried faeces for many months. Iodophores and formaldehyde are effective disinfecting agents, phenolics are less so. Intercurrent salmonellosis and, perhaps, other infections may be predisposing factors.

Signs:

- Respiratory signs.
- Greenish-yellow diarrhoea.
- Depression.
- Weakness.
- Inappetance.
- Weight loss.
- Nasal discharge.
- Conjunctivitis.
- Occasional transient ataxia in pigeons.
- Production drops in naive laying flocks

Post-mortem lesions:

- Vascular congestion.
- Wasting.
- Fibrinous pericarditis.
- Airsacculitis.
- Perihepatitis.
- Spleen enlarged and congested, may rupture in pigeons.
- Necrotic foci in liver.
- Fibrinous pneumonia.
- Congested lungs and air sacs in the turkey.

Diagnosis

- History, signs, lesions. Intracytoplasmic inclusions are helpful but confirmation requires demonstration of causal organisms (Giemsa stain, IFA).

  Serology: complement fixation, Elisa and gel diffusion.

  Differentiate from Duck viral hepatitis, Duck septicaemia.

Treatment

- Tetracycline (200-800 ppm in feed for 3-4 weeks) and/or quinolone medication and supervised slaughter.

Prevention

- Biosecurity, exclusion of wild birds.

  Live and inactivated vaccines are protective although the former result in carriers and the latter require several applications.

Chondrodystrophy, Slipped Tendon or Perosis

Caused by deficiency of manganese, choline, zinc, either singly or in combination (although deficiencies of pyridoxine, biotin, folic acid, niacin may also be involved).

This condition is seen in chickens, ducks and turkeys. In turkeys it may be an inherited deficiency of galactosamine.

Signs:

- Short legs.
- Lameness.
- Distortion of hock.
- Slipping of Achilles tendon (or perosis).
- Malposition of leg distal to hock.
- In embryos parrot beak, shortened bones.

Post-mortem lesions:

- Shortening and thickening of long bones.
- Tibia and metatarsus bowed.
- Shallow trochlea.
- Lateral slipping of tendon.

Diagnosis

- Lesions, analysis of feed.

  Differentiate from twisted leg, infectious synovitis, rickets, infectious arthritis, ruptured ligaments.

Treatment

- For flock proceed as for prevention, no value to affected bird.

Prevention

- Addition of manganese, choline, vitamins, correct mineral balance.

Coccidiosis of Turkeys

Infection of turkeys with *Eimeria* spp. This disease is not very common in commercially reared turkeys though most turkey growers receive preventative medication for at least part of their lives. Five species of *Eimeria* have been identified that cause lesions in turkeys, of which two are associated with significant disease effects.

E. meleagrimitis affects the upper small intestine, while *E. adenoides* affects the caecae and rectum. *E. gallopavonis* and *E. meleagridis* affect the lower small intestine rectum and caecae, while *E. dispersa* is found in the small intestine.

Signs:

- Huddling.
- Weight loss.
- Depression.
- Watery diarrhoea that may occasionally be blood stained or contain clumps of mucus or shed mucosa.
- Tucked appearance, ruffled feathers.

Post-mortem lesions:

- The affected area of intestine shows thickening of the wall and dilation. The contents may be haemorrhagic or be watery with white material shed from the mucosa.

Diagnosis

- Signs, lesions, microscopic exam of scrapings (oocysts, gamonts). Differentiate from necrotic enteritis.

Treatment

- Toltrazuril, Sulphonamides (e.g. Sulphaquinoxaline), Amprolium.

Prevention

- The ionophore coccidiostats lasalocid and monensin are routinely used in turkey growers, typically to 12 weeks of age. Diclazuril is also used for this purpose. Dosage levels of ionophores may be critical to efficacy and safety. Exposure of previously unmedicated birds to these compounds can cause toxicity. Salinomycin is toxic for turkeys even at very low doses. Avoid use of tiamulin in ionophore treated birds.

Coccidiosis, Caecal, E tenella

This was at one time the commonest type of coccidiosis and is certainly the most easily diagnosed. It is caused by *Eimeria tenella* and results in lesions in the caecum of chickens worldwide.

Morbidity is 10-40% and mortality up to 50%. Transmission as for *E. mitis*.

Signs:

- Depression.
- Ruffled feathers.
- Closed eyes.
- Inappetance.
- Diarrhoea, blood in faeces.
- Production less affected than in some of the other forms of coccidiosis.

Post-mortem lesions:

- Petechiae.
- Thickening, ecchymoses, of caecal mucosa.
- Accumulation of varying quantities of blood and caseous necrotic material in the caecum.

Diagnosis

- Signs, lesions, microscopic examination of scrapings. Differentiate from ulcerative enteritis, histomonosis.

Treatment

- Toltrazuril, Sulphonamides, Amprolium, Vitamins A and K in feed or water.

Prevention

- Coccidiostats in feed, vaccination by controlled exposure, hygiene. *E. tenella* is more common when 'straight' ionophore programmes are used. Shuttle programmes with chemicals in the starter diet usually improve control. In some markets the organic arsenical compound 3-Nitro is used as an aid in the control of caecal coccidiosis. Vaccines are used mainly in breeders but increasingly in broilers. Recovered birds have good immunity to the same parasite.

Coccidiosis, E mitis

This condition of chickens, seen worldwide, is caused by the protozoan parasite *Eimeria mitis*, which colonises the small intestine. The infective agent is found in litter, faeces and on fomites and birds are infected by the oral route with an incubation period of 2-5 days. The disease occurring is proportional to the amount of infective agent ingested. The parasite is moderately resistant in the environment and highly resistant to conventional disinfectants. Predisposing factors include exposure to faeces and litter conditions that favour development of the parasite (temperature, humidity).

Signs:

- Reduced feed conversion efficiency and weight gain. May predispose to wet litter, secondary bacterial enteritis.

Post-mortem lesions:

- The lesions are minimal and located in the lower small intestine (ileum) which tends to be pale and flaccid with scattered petechiae.

Diagnosis

- Mild lesions, identification of typical small round oocysts and other stages in fresh scrapings from the small intestine.

Treatment

- Not usually treated but susceptible to the products used for other forms of intestinal coccidiosis.

Prevention

- Normally controlled by anticoccidials in feed. May be included in vaccines.

Coccidiosis, E praecox

Infection of chickens with the protozoan parasite *Eimeria praecox* is spread in exactly the same as that with *E. mitis* but is practically non-pathogenic.

Signs:

- Normally asymptomatic but may cause reduced feed efficiency and reduced weight gain, and predispose to other intestinal conditions.

Post-mortem lesions:

- Minimal, but usually excess liquid and mucus in the duodenal loop.
- Severe infection can cause dehydration through excessive fluid loss.
- The cells of the sides of the villi (not tips) are usually parasitised.

Diagnosis

- Identification of characteristic slightly ovoid oocysts in the duodenum in the absence of *E. acervulina* lesions. It has a very short pre-patent period (c. 80 hours).

Treatment

- Not usually treated but susceptible to the products used for other forms of intestinal coccidiosis.

Prevention

- Normally controlled by anticoccidials in feed. Not usually included in vaccines.

Coccidiosis, Ileorectal, E brunetti

A relatively rare form of coccidiosis affecting chickens worldwide caused by *Eimeria brunetti*. Of moderate to high pathogenicity, it is found in the terminal ileum, caecum and rectum. Morbidity and mortality are variable.

Signs:

- Depression.
- Ruffled feathers.
- Closed eyes.
- Inappetance.
- Poor production.
- Diarrhoea, blood in faeces.

Post-mortem lesions:

- Petechiae and thickening of the distal third or more of intestine, extending into caecal tonsils.
- Severe necrotising enteritis.
- Oocysts in caecum and rectum.

Diagnosis

- Signs, lesions, microscopic examination of scrapings. Differentiate from ulcerative enteritis, caecal coccidiosis.

Treatment

- Toltrazuril, Sulphonamides, Amprolium, Vitamins A and K in feed or water.

Prevention

- Coccidiostats in feed, vaccination by controlled exposure, hygiene. This species is not usually included in vaccines for broilers. There is good immunity to the same parasite in recovered birds.

Coccidiosis, Intestinal, of Ducks and Geese

Young ducks and geese may suffer from *Eimeria* spp infection. In the goose *E. anseris* is the most important, while in ducks *Tyzzeria perniciosa* is most pathogenic. *Tyzerria* has eight sporocysts in each oocyst, compared to four per oocyst for *Eimeria*. Coccidiosis occurs only very rarely in commercially reared ducks in the UK.

Signs:

- Sudden death.
- Depression.
- Blood-stained vent.
- Tucked appearance.

Post-mortem lesions:

- Massive haemorrhage in upper small intestine.

Diagnosis

- Signs, lesions, microscopic examination of scrapings (usually few or no oocysts, large number of merozoites). Differentiate from Duck viral hepatitis, Duck viral enteritis, anatipestifer.

Treatment

- Sulphonamides (e.g. Sulphadimidine 30-600gm/100 birds/day, 3 days on, 2 days off, 3 days on), Amprolium, Vitamins A and K in feed or water.

Prevention

- If required coccidiostats could be used in feed, however this is not routinely practised. Hygiene.

Coccidiosis, Kidney

A disease of geese caused by *Eimeria truncata* that can cause high mortality in geese of 3-12 weeks of age, it can also infect Barbary ducks and swans.

Signs:

- Depression.
- Weakness.
- Diarrhoea - faeces tend to be whitish.
- Reduced feed intake.

Post-mortem lesions:

- Enlarged kidneys.
- Kidneys light grey to greyish pink.
- Tiny white foci and petechiae in the kidneys.

Diagnosis

- Lesions, presence of coccidial stages in fresh scrapings of kidney lesions.

Treatment

- Controlled trials of treatments have not been published.

Prevention

- Good Hygiene.

Coccidiosis, Mid-intestinal, E maxima

One of the more common forms of coccidiosis in commercial broilers. Because of the area of intestine affected it tends to have a significant effect on productivity and susceptibility to necrotic enteritis. Caused by *Eimeria maxima*, of moderate to high pathogenicity it is seen worldwide. Morbidity and mortality are variable.

Signs:

- Depression.
- Ruffled feathers.
- Closed eyes.
- Inappetance.
- Poor production.
- Blood or pigment in the faeces.
- Depigmentation of skin and plasma is especially evident in this form of cocccidiosis and this is commercially important in some markets.

Post-mortem lesions:

- Petechiae and thickening of middle third of intestine.
- Poor absorption of nutrients/pigments.
- Mild to severe enteritis, contents often orange in colour, mucosa tends to be pinker than normal.
- This infection is often associated with *E. acervulina* coccidiosis and there may be large numbers of characteristic oocysts in smears.

Diagnosis

- Signs, lesions, microscopic examination of scrapings. Differentiate from necrotic enteritis, non-specific enteritis.

Treatment

- Sulphonamides, Amprolium, Vitamins A and K in feed or water.

Prevention

Coccidiostats in feed, vaccination, hygiene. This is one of the less immunogenic species, commercial vaccines commonly contain more than one strain of *E. maxima.*

Coccidiosis, Mid-intestinal, E necatrix

A highly pathogenic form of coccidiosis, caused by *Eimeria necatrix*, in which the parasite is present in the small intestine and in the caecum. It occurs in chickens worldwide and has variable morbidity but mortality is high in severely affected birds.

Signs:

- Reduced feed consumption.
- Depression.
- Ruffled feathers.
- Closed eyes.
- Inappetance.
- Poor production.
- Diarrhoea, blood in faeces.

Post-mortem lesions:

- Petechiae and thickening, of middle to posterior third or more of small intestine.
- 'Sausage-like' intestine.
- Severe necrotising enteritis.
- Schizonts seen as white spots through the serosa interspersed with petechiae. Deep scrapings necessary to show large schizonts. Oocyts in caecal scrapings.

Diagnosis

- Signs, lesions, microscopic examination of scrapings Differentiate from necrotic enteritis, other types of coccidiosis.

Treatment

- Toltrazuril, Sulphonamides, Amprolium, Vitamins A and K in feed or water.

Prevention

- Coccidiostats in feed, vaccination, hygiene. This is one of the less immunogenic species, commercial vaccines commonly contain more than one strain of *E. maxima.*

Coccidiosis, Upper Intestinal, E acervulina

This is probably the commonest cause of coccidiosis in chickens and occurs worldwide.

It is seen in layers and in broilers, both alone and in association with other species of coccidia and is caused by *Eimeria acervulina*, which is moderately pathogenic.

Morbidity is variable and mortality low or absent.

Eimeria mivati is currently considered not to be a valid species distinct from *E. acervulina*.

Signs:

- Depression.
- Ruffled feathers.
- Closed eyes.
- Inappetance.
- Poor production.
- Diarrhoea.
- Depigmentation.

Post-mortem lesions:

- Thickening, and other lesions, restricted to upper third of small intestine - the duodenum and part of the ileum.
- Petechiae.
- White spots or bands in the mucosa. In severe infections they become confluent and cause sloughing of the mucosa.
- Poor absorption of nutrients/pigments.
- A system of assessing the severity of coccidial challenge by attributing a 'score' is often used. A detailed description is beyond the scope of this book. In general terms a score of 0 indicates no lesions and a score of 4 indicates maximal severity of lesion or death. Various publications provide a photographic key to severity of lesion.

Diagnosis

- Signs, lesions, microscopic exam of scrapings. Differentiate from necrotic and non-specific enteritis.

Treatment

- Toltrazuril, Sulphonamides, Amprolium, in feed or water.

Prevention

- Coccidiostats in feed, vaccination by controlled exposure, hygiene. Immunity is quite short lived (about 30 days) in the absence of continued challenge.

Colibacillosis, Colisepticemia

Coli-septicaemia is the commonest infectious disease of farmed poultry. It is most commonly seen following upper respiratory disease (such as Infectious Bronchitis) or Mycoplasmosis.

It is frequently associated with immunosuppressive diseases such as Infectious Bursal Disease Virus (Gumboro Disease) in chickens or Haemorrhagic Enteritis in turkeys, or in young birds that are immunologically immature. It is caused by the bacterium *Escherichia coli* and is seen worldwide in chickens, turkeys, etc.

Morbidity varies, mortality is 5-20%. The infectious agent is moderately resistant in the environment, but is susceptible to disinfectants and to temperatures of 80°C.

Infection is by the oral or inhalation routes, and via shell membranes/yolk/navel, water, fomites, with an incubation period of 3-5 days. Poor navel healing, mucosal damage due to viral infections and immunosuppression are predisposing factors.

Signs:

- Respiratory signs, coughing, sneesing.
- Snick.
- Dejection.
- Reduced appetite.
- Poor growth.
- Omphalitis.

Post-mortem lesions:

- Airsacculitis.
- Pericarditis.
- Perihepatitis.
- Swollen liver and spleen.
- Peritonitis.
- Salpingitis.
- Omphalitis.

- Synovitis.
- Arthritis.
- Enteritis.
- Granulomata in liver and spleen.
- Cellulitis over the abdomen or in the leg.
- Lesions vary from acute to chronic in the various forms of the disease.

Diagnosis

- Isolation, sero-typing, pathology. Aerobic culture yields colonies of 2-5mm on both blood and McConkey agar after 18 hours - most strains are rapidly lactose-fermenting producing brick-red colonies on McConkey agar. Differentiate from acute and chronic infections with *Salmonella* spp, other enterobacteria such as *Proteus*, as well as *Pseudomonas*, *Staphylococcus* spp., etc.

Treatment

- Amoxycillin, tetracyclines, neomycin (intestinal activity only), gentamycin or ceftiofur (where hatchery borne), potentiated sulphonamide, flouroquinolones.

Prevention

- Good hygiene in handling of hatching eggs, hatchery hygiene, good sanitation of house, feed and water. Well-nourished embryo and optimal incubation to maximise day-old viability.

Control of predisposing factors and infections (usually by vaccination).

Immunity is not well documented though both autogenous and commercial vaccines have been used.

8

Epidemic Threat from Poultry Farming

Bird Flu is a Real Pandemic Threat to Humans

As their lungs filled ... the patients became short of breath and increasingly cyanotic. After gasping for several hours they became delirious and incontinent, and many died struggling to clear their airways of a blood-tinged froth that sometimes gushed from their nose and mouth. It was a dreadful business. Isaac Starr, 3rd year medical student, University of Pennsylvania, 1918.

All the genes of all influenza viruses in the world are being maintained in aquatic birds, and periodically they transmit to other species... The 1918 viruses are still being maintained in the bird reservoir. So even though these viruses are very ancient, they still have the capacity to evolve, to acquire new genes, new hosts. The potential is still there for the catastrophe of 1918 to happen again.

Dr. Robert Webster, Influenza Expert, Present Day.

The pair of quotes presented above were used to begin and end a short essay I wrote many years ago on The 1918 Spanish Flu Pandemic. Isaac Starr's words offer powerful testimonial to the harsh lethality of a killer stain of influenza, and Dr. Robert Websters observation reminds us that it would be dangerous to regard the events of 1918 as an isolated slice of history that we can safely file away and forget.

In the early part of 1995, as I cast around in search of the disease that would ultimately prove to be the true villian of the science thriller I was planning to write, influenza was scarcely on the radar scope of anyone but virologists specialising in the field of emerging diseases. Because influenza exists in the human population in an ever-present but relatively benign form, changing just a little from year to year, we tend, quite mistakenly, to think we understand the nature of the threat that it presents.

Bird Flu cannot be eradicated, we must learn to live with it.

Despite the apparent "novelty" of a strain of bird flu that can sweep across the globe and kill not only the domesticated foul that carries the virus causing the pandemic, but also those humans unfortunate enough to get too close to the sick birds they tend, finding influenza in domesticated birds is not the least bit uncommon.

The natural reservoir for influenza virus is in fact the intestines of water fowl. In particular, ducks habour the virus, and suffer no ill effects from doing so. Because these migratory birds are always on the move, and because they shed virus in the waterways that they inhabit, influenza virus travels widely and is easily transmitted from bird to bird.

Because of this, domesticated birds, like the turkey and the chicken, which cross paths with the virus, become infected and habor the potential to transmit the virus to humans (for example, virus secreted onto the surface of eggs could end up infecting egg handlers—for this reason, if bird handlers notice any of the following symptons in their birds they should immediately notify Federal or State animal health officials: sudden death; lack of energy and appetite; decreased egg production; soft-shelled or misshapen eggs; swelling; purple discoloration; nasal discharge; coughing, sneezing; lack of coordination and diarrhea).

Often an intermediate host, the pig, is involved in the cross-species transmission of influenza. When this happens, the result is referred to as Swine Flu. Even so, the source of the virus is generally aquatic birds.

Most strains of bird flu *do not* cross species and end up infecting humans. However the H5N1 strain that has in the last decade been known to infect humans, and kill them in 5 out of every 10 cases, clearly has surmounted the first of two barriers that must be cleared before a bird flu pandemic can take hold in the human population. The second barrier will be surmounted when this strain acquires the ability to jump from one human to another. This may never happen, and hopefully will not. But based on the history of pandemic flu, it seems more likely that this is just a matter of time. Whether this takes one year or twenty is just impossible to say. The World Health Organisation has estimated that a human-to-human strain of bird flu could result in an influenza pandemic with the potential to cause between 2 and 7 million fatalities worldwide. Considering the underwhelming state of health services offered by the majority of countries in the world today, these numbers do not appear to be overhyped estimates of the potential lethality of a bird flu pandemic. Should a bird flu pandemic break out with a lethality approaching that of the 1918 Spanish Flu pandemic, it is probably fair to state that the WHO prediction is likely to fall far short of the actual death toll.

Emergence of a Bird Flu Pandemic?

The current candidate for the cause of a bird flu pandemic that crosses into the human population is the H5N1 strain. As discussed in my essay on The 1918 Spanish Flu Pandemic, and the Emerging Bird Flu Pandemic, this strain was first detected in Hong Kong in 1997, and in the time since then has spread halfway across the globe despite strenuous efforts to eradicate it in poultry stocks that have become infected. This movement of the virus, which requires transmission from one (bird) host to another via shared waterways, is slow in comparison to the movement of an airborne strain which can be transmitted from one human host to another in the particulate matter exchanged during coughing and sneezing events. The pandemic of 1918 spread across the globe in about six months, despite the dependence of that time on ship borne travel. In an age of international travel, this encircling time could be reduced to a matter of days. Once it begins, the likelihood that it can be stopped in its tracks is small.

Life during a Bird Flu Pandemic

One can never predict with any accuracy the way a natural catastrophe will unfold. But some speculation on life during the time of a bird flu pandemic seems worthwhile. Things we can say for certain are the following:

No country will be fully prepared for the event. This stems from the fact that no government is ever prepared for any natural catastrophe, as such preparations are prohibitively expensive. This means the bulk of the population will not have access to medical treatment when they need it. Families will be required to look after themselves. Only the patients identified in the very first stages of a bird flu pandemic are likely to receive treatment.

Medical supplies for combating a bird flu pandemic in the human population will be quickly exhausted. Even now, access to such antivirals as Tamiflu is poor in most parts of the world. (In late 2005 the makers of Tamiflu stepped up production to 10 times the normal level in response to worldwide demand for the drug in anticipation of an impending pandemic).

Likewise, flu vaccines, which require a long lead time for preparation (six months or more) and can easily be "spoiled" due to inadequate preparation, will not be available in anything like a sufficient supply in the foreseeable future.

Social disruption will be significant. When workers are faced with the decision of weighing the safety of family members against the worth of their job, you can bet the absentee rate will skyrocket during a pandemic. When the loss of workers begins to affect the delivery of food supplies, power, and water, we are suddenly looking at a different kind of emergency scenario—one reminiscent of the scenes that played out in New Orleans after Hurricane Katrina struck in 2005. When the services we take for granted every day suddenly disappear, people become desparate and things go to hell rather quickly. The break down of social order may be the most frightening prospect that an emerging bird flu pandemic could bring about.

Preparing for a Bird Flu Pandemic

It is tempting for all of us, the author included, to stick our heads in the sand and hope that the news stories about the possibly of an impending *bird flu pandemic* are mostly hype. But this is about as wise as living in Florida and hoping you have seen the last hurricane, or living in Los Angeles and assuming that you'll not be hit by another earthquake. Really it is just a matter of time, though no one knows how much.

There are some not so difficult things you can do to prepare, things that will also help prepare you for other emergency events. These seems practical to me:

Stock up on water. You can never have too much of this in storage, and it is cheap. Have at least a 3 month supply in stock at any time (6 months if practical), and rotate your supply so that you use up the oldest bottles at the rate you replace them with new ones. This way you will never have to worry about filtering existing water supplies if it comes to water shortages in your area.

Stock up on food supplies. This one is a little more difficult, but if you are lucky enough to have one of those huge freezers in your home, fill it up. Rotate your food, so that it is never more than 3 (or 6) months old. Be sure to buy only the foods you normally buy, as you'll be eating them regardless of whether or not any disaster strikes. Dry foods can be stored separately, but people tend too forget about them and only check their supplies in the advent of an emergency. So be prepared to toss these after a year when they are no longer edible.

Purchase some quantity of air filtration masks which, in the event of a true bird flu pandemic will be regarded as indispensible when you need to leave the house and mix with the general populace. As regards what type of mask you should obtain if you are inclined to purchase a box or two (not a bad idea as you won't be able to do it if the time comes that you need to wear them) the Centres for Disease Control and Prevention advise health workers to use a "fit-tested respirator, at least as protective as a National Institute of Occupational Safety and Health (NIOSH)-approved N-95 filtering facepiece (i.e., disposable) respirator" as one of the possible precautions against airborne infection when working with patients with known or suspected SARS or Bird Flu. 3M manufactures such masks, and the latest variety with the "Cool Flow" exhalation valve for increased comfort during extended wearing periods seems to be the mask of choice.

For other possible preparedness items you might check out a Distributor of Flu Preparedness Items (naturally enough!).

If it comes down to it, and an infectious form of bird flu arrives in your country, one of the most important things you can do to avoid infection is practice safe hygiene. This means that you go into obsessive compulsive mode and assume that everything you touch has the potential to be carrying virus. It's somewhat pointless to assume this of the contents of your own home (unless you have a stricken person on site) but once you leave the house, it is a good operating assumption. This means under no circumstances do you touch your hands to your face while in public. Children, and adults, who have the habit of putting their fingers in their eyes and noses are *extremely susceptible* to infection in such an environment. When you return to your home, clothing goes into the laundry immediately and you wash your hands thoroughly with soap to rinse away any virus you might have come in contact with. You always use a face mask in public if you have access to one. Otherwise you make one. Common sense suggests that anything that inhibits the likelihood of inhaled aerosolised influenza is better than nothing. But a proper face mask is designed to be efficient at this task. These are designed to be used *once only* and discarded. If you are a vendor, think about how you might change the operational nature of your business to stay afloat during a time of crisis. What can you do to reduce the likelihood that your employees will desert you? For instance, if you sell clothing you are very unlikely to receive my business if I must enter an enclosed space (a building) in order to purchase something. At such a time I do not want to breathe the air of a confined space. But if your stock was accessible in an open-air market I would probably be much more inclined. Employees will feel the same way, particularly if the goods in question are not critical to day to day survival.

Finally, do some more reading on the subject of bird flu pandemic. The information about bird flu found on this page is by no means exhaustive. Reading resources are listed below in the Related Links section.

Could a Bird Flu Pandemic resemble the Spanish Flu Pandemic of 1918?

A lot has happened in the last decade as regards our scientific understanding of the genetic mechanisms responsible for the virulence of some strains of influenza. Much of this new understanding has been a direct result of the work of Dr. Jeffery Taubenberger and his team at the Armed Forces Institute of Pathology. Back in 1998 I interviewed Dr. Taubenberger about his effort to reconstruct the genes of the 1918 Spanish Flu virus. At the time I had forseen the day when it would be possible to reconstruct the entire virus.

In October of 2005, after 10 years of work by Taubenberger's team, it was finally reported that the virus had been re-engineered. You can read about the announcement in the PBS article entitled 1918 Spanish Flu Offers Clues About Pandemic Viruses. Of particular interest is the observation that Taubenberger's work "published in the journal Nature in October 2005, offered some parallels between the Spanish flu virus and the H5N1 strain of the bird flu slowly spreading through Asia and recently turning up in other parts of the world." The implication here is clear. If the currently circulating H5N1 bird flu virus does acquire human-to-human transmission, the estimated mortality rates offered by the World Health Organisation for a bird flu pandemic could prove to be wildy optimistic.

Influenza Vaccine

The influenza vaccine, also known as flu shot, is an annual vaccine to protect against the highly variable influenza virus. Each injected seasonal influenza vaccine contains three influenza viruses: one influenza type A subtype H3N2 virus strain, one influenza type A subtype H1N1 (seasonal) virus strain, and one influenza type B virus strain.

Purpose and Benefits of Annual Flu Vaccination

"Influenza vaccination is the most effective method for preventing influenza virus infection and its potentially severe complications."

Deadly Epidemics each Winter

An influenza epidemic emerges during flu season each winter. There are two flu seasons annually, corresponding to the occurrence of winter in opposite months in the Northern and Southern Hemispheres.

Worldwide, seasonal influenza kills an estimated 250,000 to 500,000 people each year. Tens of thousands of Americans die in a typical flu season, but there are notable variations from year to year. In 2010 the Centres for Disease Control and Prevention (CDC) in the United States changed the way it reports the 30 year estimates for deaths from influenza. Now they are reported as a range from a low of about 3,300 deaths to a high of 49,000 per year over the past 30 years.

The majority of deaths in the industrialised world occur in adults aged 65 and over. A review at the NIAID division of the NIH in 2008 concluded that "Seasonal influenza causes more than 200,000 hospitalisations and 41,000 deaths in the U.S. every year, and is the seventh leading cause of death in the U.S." The average total economic costs caused by the annual influenza outbreak in the U.S. have been estimated at over $80 billion.

The number of annual influenza-related hospitalisations is many times the number of deaths. "The high costs of hospitalising young children for influenza creates a significant economic burden in the United States, underscoring the importance of preventive flu shots for children and the people with whom they have regular contact..." In 2006 the United States began recommending influenza vaccinations for preschoolers but Canada did not follow suit until 2010, "thereby creating a natural experiment to evaluate the effect of the policy in the United States."

A Canadian study found emergency room visits significantly lower for 2- to 4 year olds in Boston than in Montreal through the period (34% fewer ER trips). Vaccination of preschoolers may have reduced their likelihood of transmission of flu to older siblings and raised the chances that their parents would vaccinate older children as well, since there were also 18 percent fewer emergency room visits by 5- to 18 year olds in Boston than Montreal during the study period.

In another six-year observational study, vaccination of children aged 6 months through 5 years was found to prevent illness in more than half.

National Advice on Flu Vaccination

In Canada, the National Advisory Committee on Immunisation, the group that advises the Public Health Agency of Canada, currently recommends that everyone aged 2 to 64 years be encouraged to receive annual influenza vaccination, and that children between the age of six and 24 months, and their household contacts, should be considered a high priority for the flu vaccine.

In the United States, "Routine influenza vaccination is recommended for all persons aged e"6 months." The sole group for whom routine influenza vaccination is still not recommended is infants less than six months of age.

Within its blanket recommendation for general vaccination, the United States, the Centres for Disease Control and Prevention (CDC) emphasizes to clinicians the special urgency of vaccination for members of certain vulnerable groups, and their caregivers:

Vaccination is especially important for people at higher risk of serious influenza complications or people who live with or care for people at higher risk for serious complications.

Benefits of Vaccination

Vaccination against influenza is also, according to research published in July 2010, thought to be important for members of high-risk groups who would be likely to suffer complications from influenza, for example pregnant women and children and teenagers from six months to 18 years of age; In expanding the new upper age limit to 18 years, the aim is to reduce both the time children and parents lose from visits to pediatricians and missing school and the need for antibiotics for complications.

An added expected benefit would be indirect — to reduce the number of influenza cases among parents and other household members, and possibly spread to the general community.

Vaccination of school-age children has a strong protective effect on the adults and elderly with whom the children are in contact.

Children born to mothers who received flu vaccination while pregnant are strongly protected from having to be hospitalised with the flu. "The effectiveness of influenza vaccine given to mothers during pregnancy in preventing hospitalisation among their infants, adjusted for potential confounders, was 91.5%."

Healthy, working adults who received influenza vaccine reported 25 percent fewer episodes of upper respiratory illness than those who received the placebo (105 vs. 140 episodes per 100 subjects, $P < 0.001$), 43 percent fewer days of sick leave from work due to upper respiratory illness (70 vs. 122 days per 100 subjects, $P = 0.001$), and 44 percent fewer visits to physicians' offices for upper respiratory illnesses (31 vs. 55 visits per 100 subjects, $P = 0.004$). The study, reported in the NEJM, estimated cost savings at $46.85 per person vaccinated, and concluded that "Vaccination against influenza has substantial health-related and economic benefits for healthy, working adults."

Influenza vaccination has been shown highly effective in health care workers, with minimal adverse effects. In a study of forty matched nursing homes, staff influenza vaccination rates were 69.9% in the vaccination arm versus 31.8% in the control arm. The vaccinated staff experienced a 42% reduction in sick leave from work (P=.03). A review of eighteen studies likewise found a strong net benefit to health care workers. The two of these eighteen studies that assessed the relationship of patient mortality relative to staff influenza vaccine uptake both found that higher rates of health care worker vaccination correlated with reduced patient deaths. An analysis of data and patient population health in New Mexico's 75 long-term care facilities nursing homes found that as vaccination rates of health care personnel with direct patient contact rose from 51 to 75 percent, the chances of a flu outbreak among patients in that facility went down by 87 percent. The New Mexico study showed that vaccinating health care personnel provided more protection to residents than vaccinating residents themselves.

In a 2010 survey of healthcare workers, 63.5% reported that they received the flu vaccine during the 2010-11 season, an increase from 61.9% reported the previous season. Health professionals with direct patient contact had higher vaccination uptake, such as physicians and dentists (84.2%) and nurse practitioners (82.6%).

Cross-protection

Annual seasonal flu vaccination may provide some level of protection against novel flu viruses. A number of studies suggest that seasonal flu vaccine may offer cross-protection, both against the H5N1-type (avian influenza) H5N1 infection and the 2009 flu pandemic (the H1N1 "swine flu.") Vaccine protection can be long-lasting. Participants who received vaccination against the swine flu in 1976 still enjoyed benefits 33 years later, exhibiting a significantly enhanced immune response to the 2009 pandemic H1N1.

History of the Flu Vaccine

Vaccines are used in both humans and nonhumans. Human vaccine is meant unless specifically identified as a veterinary, poultry or livestock vaccine.

Influenza

The first influenza pandemic was recorded in 1580; since this time, various methods have been employed to eradicate its cause. The etiological cause of influenza, the orthomyxoviridae was finally discovered by the Medical Research Council (MRC) of the United Kingdom in 1933.

Known Flu Pandemics

1889–90 — Asiatic (Russian) Flu, mortality rate said to be 0.75–1 death per 1000 possibly H2N2

1900 — Possibly H3N8

1918–20 – Spanish Flu, 500 million ill, at least 20–40 million died of H1N1

1957–58 – Asian Flu, 1 to 1.5 million died of H2N2

1968–69 – Hong Kong Flu, 3/4 to 1 million died of H3N2

2009 - Swine Flu, caused by H1N1/09, 14,286 died.

Flu Vaccine Origins and Development

In the world wide Spanish flu pandemic of 1918, "Physicians tried everything they knew, everything they had ever heard of, from the ancient art of bleeding patients, to administering oxygen, to developing new vaccines and sera (chiefly against what we now call *Hemophilus influenzae*—a name derived from the fact that it was originally considered the etiological agent—and several types of pneumococci). Only one therapeutic measure, transfusing blood from recovered patients to new victims, showed any hint of success."

In 1931, viral growth in embryonated hens' eggs was reported by Ernest William Goodpasture and colleagues at Vanderbilt University. The work was extended to growth of influenza virus by several workers, including Thomas Frances, Wilson Smith and Macfarlane Burnet, leading to the first experimental influenza vaccines. In the 1940s, the US military developed the first approved inactivated vaccines for influenza, which were used in the Second World War. Greater advances were made in vaccinology and immunology, and vaccines became safer and mass-produced. Today, thanks to the advances of molecular technology, we are on the verge of making influenza vaccines through the genetic manipulation of influenza genes.

Flu Vaccine Acceptance

According to the CDC: "Influenza vaccination is the primary method for preventing influenza and its severe complications. Vaccination is associated with reductions in influenza-related respiratory illness and physician visits among all age groups, hospitalisation and death among persons at high risk, otitis media among children, and work absenteeism among adults. Although influenza vaccination levels increased substantially during the 1990s, further improvements in vaccine coverage levels are needed".

The current egg-based technology for producing influenza vaccine was created in the 1950s. In the U.S. swine flu scare of 1976, President Gerald Ford was confronted with a potential swine flu pandemic. The vaccination program was rushed, yet plagued by delays and public relations problems. Meanwhile, maximum military containment efforts succeeded unexpectedly in confining the new strain to the single army base where it had originated. On that base a number of soldiers fell severely ill, but only one died. The program was cancelled, after about 24% of the population had received vaccinations. An excess in deaths of twenty-five over normal annual levels as well as 400 excess hospitalisations, both from Guillain-Barré syndrome, were estimated to have occurred from the vaccination program itself, illustrating that vaccine itself is not free of risks. The result has been cited to stoke lingering doubts about vaccination. In the end, however, even the maligned 1976 vaccine may have saved lives. A 2010 study found a significantly enhanced immune response against the 2009 pandemic H1N1 in study participants who had received vaccination against the swine flu in 1976.

Current Status

Influenza research includes molecular virology, molecular evolution, pathogenesis, host immune responses, genomics, and epidemiology. These help in developing influenza countermeasures such as vaccines, therapies and diagnostic tools. Improved influenza countermeasures require basic research on how viruses enter cells, replicate, mutate, evolve into new strains and induce an immune response. The Influenza Genome Sequencing Project is creating a library of influenza sequences that will help us understand what makes one strain more lethal than another, what genetic determinants most affect immunogenicity, and how the virus evolves over time. Solutions to limitations in current vaccine methods are being researched.

The rapid development, production, and distribution of pandemic influenza vaccines could potentially save millions of lives during an influenza pandemic. Due to the short time frame between identification of a pandemic strain and need for vaccination, researchers are looking at novel technologies for vaccine production that could provide better "real-time" access and be produced more affordably, thereby increasing access for people living in low- and moderate-income countries, where an influenza pandemic may likely originate, such as live attenuated (egg-based or cell-based) technology and recombinant technologies (proteins and virus-like particles).

As of July 2009, more than 70 known clinical trials have been completed or are ongoing for pandemic influenza vaccines. In September 2009, the US Food and Drug Administration approved four vaccines against the 2009 H1N1 influenza virus (the current pandemic strain), and expected the initial vaccine lots to be available within the following month.

Prospects for Universal Flu Vaccines

Many groups world wide are working on a universal flu vaccine that will not need changing each year, as the sector has been viewed as "increasingly hot". Companies pursuing the vaccine as of 2009 and 2010 include BiondVax, Theraclone, Dynavax Technologies Corporation, VaxInnate, Crucell NV, Inovio Pharmaceuticals, and Immune Targeting Systems (ITS)

In 2008 Acambis announced work on a universal flu vaccine (ACAM-FLU-ATM) based on the less variable M2 protein component of the flu virus shell.

In 2009, the Wistar Institute received a patent for using "a variety of peptides" in a flu vaccine, and announced it was seeking a corporate partner.

In 2010, the National Institute of Allergy and Infectious Diseases (NIAID) of the U.S. NIH announced a breakthrough; the effort targets the stem, which mutates less often than the head of the virus.

DNA vaccines such as VGX-3400X (aimed at multiple H5N1 strains) contain DNA fragments (plasmids). Inovios SynCon DNA vaccines include H5N1 and H1N1 subtypes.

In July 2011, F16 researchers created an antibody, which targets a protein found on the surface of all influenza A viruses called haemagglutinin.

F16 is the only known antibody that treats all 16 subtypes of the influenza A virus and might be the lynchpin for a universal influenza vaccine.

Other Vaccines are Polypeptide Based

Some universal flu vaccines have started early stage clinical trials.

BiondVax are targeting the less variable stalk of the haemagglutinin molecule with Multimeric-001. This is aimed at type A (inc H1N1) and Type B influenza and has started a phase IIa study.

Dynavax have developed a vaccine N8295 based on two highly conserved antigens NP and M2e and their TLR9 agonist, and started clinical trials in June 2010.

ITS's fp01 includes 6 peptide antigens to highly conserved segments of the PA, PB1, PB2, NP & M1 proteins, and has started phase I trials.

Based on the results of animal studies, a universal flu vaccine may use a two-step vaccination strategy — priming with a DNA-based HA vaccine followed by a second dose with an inactivated, attenuated, or adenovirus-vector–based vaccine.

Some people given a 2009 H1N1 flu vaccine have developed broadly protective antibodies which has raised hopes for a universal flu vaccine.

Clinical Trials of Vaccines

A vaccine is assessed by the reduction of the risk of disease that is produced by vaccination, the vaccine's *efficacy*. In contrast, in the field, the *effectiveness* of a vaccine is the practical reduction in risk for an individual when they are vaccinated under real-world conditions. Measuring efficacy of influenza vaccines is relatively simple, as the immune response produced by the vaccine can be assessed in animal models, or the amount of antibody produced in vaccinated people can be measured, or most rigorously, by immunising adult volunteers and then challenging with virulent influenza virus. In studies such as these, influenza vaccines showed high efficacy and produced a protective immune response. For ethical reasons, such challenge studies cannot be performed in the population most at risk from influenza – the elderly and young children. However, studies on the effectiveness of flu vaccines in the real world are uniquely difficult. The vaccine may not be matched to the viruses in circulation that year; virus prevalence varies widely between years, and influenza is often confused with other influenza-like illnesses.

Nevertheless, multiple clinical trials of both live and inactivated influenza vaccines against seasonal influenza have been performed and their results pooled and analysed in several recent meta-analyses. Studies on live vaccines have very limited data, but these preparations may be more effective than inactivated vaccines. The meta-analyses examined the efficacy and effectiveness of inactivated vaccines against seasonal influenza in adults, children, and the elderly. In adults, vaccines show high efficacy against the targeted strains, but low effectiveness overall, so the benefits of vaccination are small, with a one-quarter reduction in risk of contracting influenza but no significant effect on the rate of hospitalisation. However, the risk of serious complications from influenza is small in adults, so unless the effect from vaccination is large it might not have been detected. In children, vaccines again showed high efficacy, but low effectiveness in preventing "flu-like illness". In children under two the data are extremely limited, but vaccination appeared to confer no measurable benefit. In the elderly, vaccination does not reduce the frequency of influenza, but seems to reduce pneumonia, hospital admission and deaths from influenza or pneumonia. However, the current data on the effectiveness of influenza vaccines in the elderly may be unreliable, due to high levels of selection bias.

Overall, the benefit of influenza vaccination is clear in the elderly and vaccination of children may be beneficial. Routine vaccination of adults is not predicted to produce significant improvements in public health. The apparent contradiction between vaccines with high efficacy, but low effectiveness, may reflect the difficulty in diagnosing influenza under clinical conditions and the large number of strains circulating in the population. In contrast, during an influenza pandemic, where a single strain of virus is responsible for illnesses, an effective vaccine could produce a large decrease in the number of cases and be highly effective in controlling an epidemic. However, such a vaccine would have to be produced and distributed rapidly to have maximum effect.

Effectiveness of Vaccine

The CDC reports that studies demonstrate that vaccination is a cost-effective counter-measure to seasonal outbreaks of influenza. However it is not perfect. A study led by Dr. David K. Shay in February, 2008 reported that:

> *"full immunisation against flu provided about a 75 percent effectiveness rate in preventing hospitalisations from influenza complications in the 2005-6 and 2006-7 influenza seasons."*

Influenza vaccine has been demonstrated to prevent disease and death, both in numerous controlled studies and in painstaking scientific reviews of these studies. However the rigor of the science of these studies has also been criticized. A 2006 Cochrane review of influenza vaccination in the elderly stated "The apparent high effectiveness of the vaccines in preventing death from all causes may reflect a baseline imbalance in health status and other systematic differences in the two groups of participants. In one observational study, a sharply lower risk of death or hospitalisation for pneumonia was seen in vaccinated persons:

Results: The relative risk of death for vaccinated persons compared with unvaccinated persons was 0.39 [95% confidence interval (95% CI), 0.33-0.47] before influenza season, 0.56 (0.52-0.61) during influenza season, and 0.74 (0.67-0.80) after influenza season. The relative risk of pneumonia hospitalisation was 0.72 (0.59-0.89) before, 0.82 (0.75-0.89) during, and 0.95 (0.85-1.07) after influenza season. Adjustment for diagnosis code variables resulted in estimates that were further from the null, in all time periods.

A better vitality and lower pneumonia hospitalisation was thus observed in the vaccinated group in this non-randomised population study before, during, and after influenza season. This could, however, have been ascribed in whole or in part to self-selection bias, since vaccinated persons were already healthier before flu season. Some sub-populations have been assumed to benefit from vaccination in the absence of directly specific studies. For example, a 2008 Cochrane review of healthy children found "Influenza vaccines are efficacious in children older than two but little evidence is available for children under two.". The CDC in 2010, after a review of extant studies, extended its guidelines to recommend that every child over 6 months be given the influenza vaccine. Vaccines were shown effective against the influenza strains they are designed to protect against, but this translated to only a modest impact on working days lost due to influenza-like infections in a 2007 Cochrane review on influenza vaccines in healthy adults. While a 2010 Cochrane review noted that "Influenza vaccines have a modest effect in reducing influenza symptoms and working days lost" it stated found no evidence of prevention of complications, such as pneumonia, or transmission.

The group most vulnerable to non-pandemic flu, the elderly, is also the least benefitted by the vaccine, with an average efficacy rate ranging from 40-50% at age 65, and only 15-30% past age 70. There are multiple reasons behind this steep decline in vaccine efficacy, the most common of which are the declining immunological function and frailty associated with advanced age. An influenza vaccine with four times the usual amount of antigen (Fluzone High Dose) has shown increased immune response in the elderly and has now been approved by the U.S. Food and Drug Administration (FDA).

In a non-pandemic year, a person in the United States aged 50–64 is nearly ten times more likely to die an influenza-associated death than a younger person, and a person over age 65 is over ten times more likely to die an influenza-associated death than the 50–64 age group. Vaccination of those over age 65 reduces influenza-associated death by about 50%. However, it is unlikely that the vaccine completely explains the results since elderly people who get vaccinated are probably more healthy and health-conscious than those who do not. Elderly participants randomised to a high-dose group (60 micrograms) had antibody levels 44 to 79 percent higher than did those who received the normal dose of vaccine. Elderly volunteers receiving the higher dose were more likely to achieve protective levels of antibody.

As mortality is also high among infants who contract influenza, the household contacts and caregivers of infants should be vaccinated to reduce the risk of passing an influenza infection to the infant. Data from the years when Japan required annual flu vaccinations for school-aged children indicate that vaccinating children the group most likely to catch and spread the disease—has a strikingly positive effect on reducing mortality among older people: one life saved for every 420 children who received the flu vaccine.

This may be due to herd immunity or to direct causes, such as individual older people not being exposed to influenza. For example, retired grandparents often risk infection by caring for their sick grandchildren in households where the parents can't take time off work or are sick themselves.

In most years (16 of the 19 years before 2007), the flu vaccine strains have been a good match for the circulating strains. In other flu seasons like that of 2007/2008, the match was less useful. But even a mismatched vaccine can often provide some protection:

Antibodies made in response to vaccination with one strain of influenza viruses can provide protection against different, but related strains. A less than ideal match may result in reduced vaccine effectiveness against the variant viruses, but it still can provide enough protection to prevent or lessen illness severity and prevent flu-related complications.

In addition, it's important to remember that the influenza vaccine contains three virus strains so the vaccine can also protect against the other two viruses. For these reasons, even during seasons when there is a less than ideal match, CDC continues to recommend influenza vaccination. This is particularly important for people at high risk for serious flu complications and their close contacts.

Comparing Flu Shot to Nasal Spray

Flu vaccines are available either as:

TIV (flu shot (injection) of trivalent (three strains; usually A/H1N1, A/H3N2, and B) inactivated (killed) vaccine) or

LAIV (nasal spray (mist) of live attenuated influenza vaccine.)

TIV works by putting into the bloodstream those parts of three strains of flu virus that the body uses to create antibodies; while LAIV works by inoculating against those same three strains that have been genetically modified to minimise symptoms of illness.

LAIV is not recommended for individuals under age 2 or over age 50, but might be comparatively more effective among children over age 2.

A military study on military personnel showed that flu shots yielded less illness than nasal spray. Based on one of the largest head-to-head studies comparing LAIV and TIV (which was conducted by the U.S. Armed Forces Surveillance Centre on military personnel who were stationed in the United States during three flu seasons from 2004 through 2007), investigators concluded that: "It may be prudent to use TIV in patients who were vaccinated at least once in the past 2 years but LAIV against pandemic strains may be more protective than inactivated vaccines, because the population will probably lack preexisting immunity."

High-dose Vaccine

A high-dose vaccine (Fluzone High-Dose) 4x the strength of standard flu vaccine was approved by the FDA in late 2009. This vaccine is intended for people 65 and over, who typically have weakened immune response due to normal aging.

The vaccine produces a greater immune response than standard vaccine, but it is not yet known whether it provides greater protection against flu. Study results are expected in 2012. CDC recommends the high-dose vaccine for people 65 and over but expresses no preference between it and standard vaccine.

Vaccination Recommendations

Various public health organisations, including the World Health Organisation, have recommended that yearly influenza vaccination be routinely offered to patients at risk of complications of influenza and those individuals who live with or care for high-risk individuals, including:

- the elderly (UK recommendation is those aged 65 or above)
- patients with chronic lung diseases (asthma, COPD, etc.)
- patients with chronic heart diseases (congenital heart disease, chronic heart failure, ischaemic heart disease)
- patients with chronic liver diseases (including cirrhosis)
- patients with chronic renal diseases (such as the nephrotic syndrome)
- patients who are immunosuppressed (those with HIV or who are receiving drugs to suppress the immune system such as chemotherapy and long-term steroids) and their household contacts
- people who live together in large numbers in an environment where influenza can spread rapidly, such as prisons, nursing homes, schools, and dormitories
- healthcare workers (both to prevent sickness and to prevent spread to patients)
- pregnant women. However, a 2009 review concluded that there was insufficient evidence to recommend routine use of trivalent influenza vaccine during the first trimester of pregnancy.
- children from ages six months to two years.

Both types of flu vaccines are contraindicated for those with severe allergies to egg proteins and people with a history of Guillain-

Barré syndrome. Public Health Law Research, an independent organisation, published in 2009 several evidence briefs summarising the research assessing the effect of specific laws and policy on public health. There is sufficient evidence supporting the effectiveness of requiring vaccinations as a condition for attending child care facilities and schools. There is insufficient evidence to assess the effectiveness of requiring vaccinations as a condition for specified jobs as a means of reducing incidence of specific diseases among particularly vulnerable populations.

There is strong evidence supporting the effectiveness of standing orders which allow healthcare workers without prescription authority to administer vaccines under defined circumstances as a public health intervention aimed at increasing vaccination rates.

Side effects

- Side effects of the inactivated/dead flu vaccine injection include:
- mild soreness, redness, and swelling where the shot was given
- fever
- aches
- These problems usually begin soon after the injection, and last 1–2 days.
- Side effects of the activated/live/LAIV flu nasal spray vaccine:
- Some children and adolescents 2–17 years of age have reported:
- runny nose, nasal congestion or cough
- fever
- headache and muscle aches
- wheesing
- abdominal pain or occasional vomiting or diarrhea
- Some adults 18–49 years of age have reported:
- runny nose or nasal congestion
- sore throat
- cough, chills, tiredness/weakness
- headache
- More severe, but very rare side effects include:
- life-threatening allergic reaction.

Some injection-based flu vaccines intended for adults in the United States contain thiomersal (also known as thimerosal). Despite some controversy in the media, the World Health Organisation has concluded that there is no evidence of toxicity from thiomersal in vaccines and no reason on grounds of safety to change to more-expensive single-dose administration.

Although Guillain-Barré syndrome had been feared as a complication of vaccination, the CDC states that most studies on modern influenza vaccines have seen no link with Guillain-Barré.

A review has concluded that the 2009 H1N1 ("swine flu") vaccine has a safety profile similar to that of seasonal vaccine. Although one review gives an incidence of about one case per million vaccinations, a large study in China, reported in the NEJM covering close to 100 million doses of vaccine against the 2009 H1N1 "swine" flu found only eleven cases of Guillain-Barre syndrome, (0.1%) total incidence in persons vaccinated, actually lower than the normal rate of the disease in China, and no other notable side effects; "The risk-benefit ratio, which is what vaccines and everything in medicine is about, is overwhelmingly in favour of vaccination." Getting infected by influenza itself increases both the risk of death (up to 1 in 10,000) and increases the risk of developing Guillain-Barré syndrome to a much higher level than the highest level of suspected vaccine involvement (approx. 10 times higher by recent estimates).

Flu Vaccine Manufacturing

Flu vaccine is usually grown in fertilised chicken eggs. In February preceding each fall's flu season (in the Northern hemisphere), three strains of flu are selected and chicken eggs inoculated.

As of November 2007, both the conventional injection and the nasal spray are manufactured using chicken eggs. The European Union has also approved Optaflu, a vaccine produced by Novartis using vats of animal cells. This technique is expected to be more scalable and avoid problems with eggs, such as allergic reactions and incompatibility with strains that affect avians like chickens. A DNA-based vaccination, which is hoped to be even faster to manufacture, is currently in clinical trials, but has not yet been proven safe and effective. Research continues into the idea of a "universal" influenza vaccine (but no vaccine candidates have been announced) which would not need to be tailored to work on particular strains, but would be effective against a broad variety of influenza viruses.

In a 2007 report, the current global capacity of approximately 826 million seasonal influenza vaccine doses (inactivated and live) was double the current production of 413 million doses. In an aggressive scenario of producing pandemic influenza vaccines by 2013, only 2.8 billion courses could be produced in a six-month time frame. If all high- and upper-middle-income countries sought vaccines for their entire populations in a pandemic, nearly 2 billion courses would be required. If China pursued this goal as well, more than 3 billion courses would be required to serve these populations. Vaccine research and development is ongoing to identify novel vaccine approaches that could produce much greater quantities of vaccine at a price that is affordable to the global population.

An effective method of vaccine generation that bypasses the need for eggs is the construction of "influenza virus-like particle (VLP)". VLP is a non-egg, non-mammalian cell culture-based vaccine, purified from the supernatants of Spodoptera frugiperda Sf9 insect cells following infection of baculovirus vectors encoding an expression cassette made up of only three influenza virus structural proteins, hemagglutinin (HA), neuraminidase (NA), and matrix (M1) VLPs elicit antibodies that recognise a broader panel of antigenically distinct viral isolates compared to other vaccines in the hemagglutination-inhibition (HAI) assay.

H5N1

Vaccines have been formulated against several of the avian H5N1 influenza varieties. Vaccination of poultry against the ongoing H5N1 epizootic is widespread in certain countries. Some vaccines also exist for use in humans, and others are in testing, but none have been made available to civilian populations, nor produced in quantities sufficient to protect more than a tiny fraction of the Earth's population in the event of an H5N1 pandemic.

- Three H5N1 vaccines for humans have been licensed as of June 2008:
- Sanofi Pasteur's vaccine approved by the United States in April 2007,
- GlaxoSmithKline's vaccine Pandemrix approved by the European Union in May 2008, and
- CSL Limited's vaccine approved by Australia in June 2008.
- All are produced in eggs and would require many months to be altered to a pandemic version.

H5N1 continually mutates, meaning vaccines based on current samples of avian H5N1 cannot be depended upon to work in the case of a future pandemic of H5N1. While there can be some cross-protection against related flu strains, the best protection would be from a vaccine specifically produced for any future pandemic flu virus strain. Dr. Daniel Lucey, co-director of the Biohazardous Threats and Emerging Diseases graduate program at Georgetown University, has made this point, "There is no H5N1 pandemic so there can be no pandemic vaccine." However, "pre-pandemic vaccines" have been created; are being refined and tested; and do have some promise both in furthering research and preparedness for the next pandemic. Vaccine manufacturing companies are being encouraged to increase capacity so that if a pandemic vaccine is needed, facilities will be available for rapid production of large amounts of a vaccine specific to a new pandemic strain.

Problems with H5N1 vaccine production include:

- lack of overall production capacity
- lack of surge production capacity (it is impractical to develop a system that depends on hundreds of millions of 11 day old specialised eggs on a standby basis)
- the pandemic H5N1 might be lethal to chickens.

Cell culture (cell-based) manufacturing technology can be applied to influenza vaccines as they are with most viral vaccines and thereby solve the problems associated with creating flu vaccines using chicken eggs as is currently done.:

Currently, influenza vaccine for the annual, seasonal influenza program comes from four manufacturers. However, only a single manufacturer produces the annual vaccine entirely within the U.S. Thus, if a pandemic occurred and existing U.S.-based influenza vaccine manufacturing capacity was completely diverted to producing a pandemic vaccine, supply would be severely limited. Moreover, because the annual influenza manufacturing process takes place during most of the year, the time and capacity to produce vaccine against potential pandemic viruses for a stockpile, while continuing annual influenza vaccine production, is limited. Since supply will be limited, it is critical for HHS to be able to direct vaccine distribution in accordance with predefined groups; HHS will ensure the building of capacity and will engage states in a discussion about the purchase and distribution of pandemic influenza vaccine.

Vaccine production capacity: The protective immune response generated by current influenza vaccines is largely based on viral hemagglutinin (HA) and neuraminidase (NA) antigens in the vaccine. As a consequence, the basis of influenza vaccine manufacturing is growing massive quantities of virus in order to have sufficient amounts of these protein antigens to stimulate immune responses. Influenza vaccines used in the United States and around world are manufactured by growing virus in fertilised hens' eggs, a commercial process that has been in place for decades. To achieve current vaccine production targets millions of 11-day old fertilised eggs must be available every day of production.

In the near term, further expansion of these systems will provide additional capacity for the U.S.-based production of both seasonal and pandemic vaccines, however, the surge capacity that will be needed for a pandemic response cannot be met by egg-based vaccine production alone, as it is impractical to develop a system that depends on hundreds of millions of 11-day old specialised eggs on a standby basis. In addition, because a pandemic could result from an avian influenza strain that is lethal to chickens, it is impossible to ensure that eggs will be available to produce vaccine when needed.

In contrast, cell culture manufacturing technology can be applied to influenza vaccines as they are with most viral vaccines (e.g., polio vaccine, measles-mumps-rubella vaccine, chickenpox vaccine). In this system, viruses are grown in closed systems such as bioreactors containing large numbers of cells in growth media rather than eggs. The surge capacity afforded by cell-based technology is insensitive to seasons and can be adjusted to vaccine demand, as capacity can be increased or decreased by the number of bioreactors or the volume used within a bioreactor. In addition to supporting basic research on cell-based influenza vaccine development, HHS is currently supporting a number of vaccine manufacturers in the advanced development of cell-based influenza vaccines with the goal of developing U.S.-licensed cell-based influenza vaccines produced in the United States. The US government has purchased from Sanofi Pasteur and Chiron Corporation several million doses of vaccine meant to be used in case of an influenza pandemic of H5N1 avian influenza and is conducting clinical trials with these vaccines. Researchers at the University of Pittsburgh have had success with a genetically engineered vaccine that took only a month to make and completely protected chickens from the highly pathogenic H5N1 virus.

According to the United States Department of Health & Human Services

In addition to supporting basic research on cell-based influenza vaccine development, HHS is currently supporting a number of vaccine manufacturers in the advanced development of cell-based influenza vaccines with the goal of developing U.S.-licensed cell-based influenza vaccines produced in the United States. Dose-sparing technologies. Current U.S.-licensed vaccines stimulate an immune response based on the quantity of HA (hemagglutinin) antigen included in the dose. Methods to stimulate a strong immune response using less HA antigen are being studied in H5N1 and H9N2 vaccine trials. These include changing the mode of delivery from intramuscular to intradermal and the addition of immune-enhancing adjuvant to the vaccine formulation. Additionally, HHS is soliciting contract proposals from manufacturers of vaccines, adjuvants, and medical devices for the development and licensure of influenza vaccines that will provide dose-sparing alternative strategies.

Chiron Corporation is now recertified and under contract with the National Institutes of Health to produce 8,000–10,000 investigational doses of Avian Flu (H5N1) vaccine. MedImmune and Aventis Pasteur are under similar contracts. The United States government hopes to obtain enough vaccine in 2006 to treat 4 million people. However, it is unclear whether this vaccine would be effective against a hypothetical mutated strain that would be easily transmitted through human populations, and the shelflife of stockpiled doses has yet to be determined.

The *New England Journal of Medicine* reported on March 30, 2006 on one of dozens of vaccine studies currently being conducted. The Treanor et al. study was on vaccine produced from the human isolate (A/Vietnam/1203/2004 H5N1) of a virulent clade 1 influenza A (H5N1) virus with the use of a plasmid rescue system, with only the hemagglutinin and neuraminidase genes expressed and administered without adjuvant. "The rest of the genes were derived from an avirulent egg-adapted influenza A/PR/8/34 strain. The hemagglutinin gene was further modified to replace six basic amino acids associated with high pathogenicity in birds at the cleavage site between hemagglutinin 1 and hemagglutinin 2. Immunogenicity was assessed by microneutralisation and hemagglutination-inhibition assays with the use of the vaccine virus, although a subgroup of samples were tested with the use of the wild-type influenza A/Vietnam/ 1203/2004 (H5N1) virus." The results of this study combined with others scheduled to be completed by spring 2007 is hoped will provide a highly immunogenic vaccine that is cross-protective against heterologous influenza strains.

On August 18, 2006. the World Health Organisation changed the H5N1 strains recommended for candidate vaccines for the first time since 2004. "The WHO's new prototype strains, prepared by reverse genetics, include three new H5N1 subclades. The hemagglutinin sequences of most of the H5N1 avian influenza viruses circulating in the past few years fall into two genetic groups, or clades. Clade 1 includes human and bird isolates from Vietnam, Thailand, and Cambodia and bird isolates from Laos and Malaysia. Clade 2 viruses were first identified in bird isolates from China, Indonesia, Japan, and South Korea before spreading westward to the Middle East, Europe, and Africa. The clade 2 viruses have been primarily responsible for human H5N1 infections that have occurred during late 2005 and 2006, according to WHO. Genetic analysis has identified six subclades of clade 2, three of which have a distinct geographic distribution and have been implicated in human infections:

- Subclade 1, Indonesia
- Subclade 2, Middle East, Europe, and Africa
- Subclade 3, China

On the basis of the three subclades, the WHO is offering companies and other groups that are interested in pandemic vaccine development these three new prototype strains:

- An A/Indonesia/2/2005-like virus
- An A/Bar headed goose/Quinghai/1A/2005-like virus
- An A/Anhui/1/2005-like virus.

Until now, researchers have been working on prepandemic vaccines for H5N1 viruses in clade 1. In March, the first clinical trial of a U.S. vaccine for H5N1 showed modest results. In May, French researchers showed somewhat better results in a clinical trial of an H5N1 vaccine that included an adjuvant. Vaccine experts aren't sure if a vaccine effective against known H5N1 viral strains would be effective against future strains. Although the new viruses will now be available for vaccine research, WHO said clinical trials using the clade 1 viruses should continue as an essential step in pandemic preparedness, because the trials yield useful information on priming, cross-reactivity, and cross-protection by vaccine viruses from different clades and subclades."

As of November 2006, the United States Department of Health and Human Services still had enough H5N1 pre-pandemic vaccine to treat about 3 million people (5.9 million full-potency doses) in spite of 0.2 million doses used for research and 1.4 million doses that have begun to lose potency (from the original 7.5 million full-potency doses purchased from Sanofi Pasteur and Chiron Corp.). The expected shelf life of seasonal flu vaccine is about a year so the fact that most of the H5N1 pre-pandemic stockpile is still good after about 2 years is considered encouraging.

Annual Reformulation of Flu Vaccine

Each year, three strains are chosen for selection in that year's flu vaccination by the WHO Global Influenza Surveillance Network. The chosen strains are the H1N1, H3N2, and Type-B strains thought most likely to cause significant human suffering in the coming season. Due to the high mutation rate of the virus a particular vaccine formulation is effective for at most about a year. The World Health Organisation coordinates the contents of the vaccine each year to contain the most likely strains of the virus to attack the next year.

"The WHO Global Influenza Surveillance Network was established in 1952. The network comprises 4 WHO Collaborating Centres (WHO CCs) and 112 institutions in 83 countries, which are recognised by WHO as WHO National Influenza Centres (NICs). These NICs collect specimens in their country, perform primary virus isolation and preliminary antigenic characterisation. They ship newly isolated strains to WHO CCs for high level antigenic and genetic analysis, the result of which forms the basis for WHO recommendations on the composition of influenza vaccine for the Northern and Southern Hemisphere each year."

The Global Influenza Surveillance Network's selection of viruses for the vaccine manufacturing process is based on its best estimate of which strains will be predominant the next year, amounting in the end to well-informed but fallible guesswork. Formal WHO recommendations first issued in 1973; beginning 1999 there have been two recommendations per year, one for the northern hemisphere (N) and the other for the southern hemisphere (S).

History of past WHO seasonal influenza vaccine composition recommendations:

2002–2003 Northern Hemisphere Winter Season

The vaccines produced for the 2002–2003 season use:

- an A/New Caledonia/20/1999-like(H1N1);
- an A/Moscow/10/1999-like(H3N2);
- a B/Hong Kong/330/2001-like viruses.

2003 Southern Hemisphere Winter Season

The composition of influenza virus vaccines for use in the 2003 Southern Hemisphere influenza season recommended by the World Health Organisation was:

- an A/New Caledonia/20/99(H1N1)-like virus
- an A/Moscow/10/99(H3N2)-like virus (The widely used vaccine strain is A/Panama/2007/99)
- a B/Hong Kong/330/2001-like virus (Some vaccine strains used were B/Shandong/7/97, B/Hong Kong/330/2001, B/Hong Kong/1434/2002).

2003–2004 Northern Hemisphere Winter Season

The production of flu vaccine requires a lead time of about six months before the season. It is possible that by flu season a strain becomes common for which the vaccine does not provide protection. In the 2003–2004 season the vaccine was produced to protect against A/Panama, A/New Caledonia, and B/Hong Kong. A new strain, A/Fujian, was discovered after production of the vaccine started and vaccination gave only partial protection against this strain.

Nature magazine reported that the Influenza Genome Sequencing Project, using phylogenetic analysis of 156 H3N2 genomes, "explains the appearance, during the 2003–2004 season, of the 'Fujian/411/2002'-like strain, for which the existing vaccine had limited effectiveness" as due to an epidemiologically significant reassortment. "Through a reassortment event, a minor clade provided the haemagglutinin gene that later became part of the dominant strain after the 2002–2003 season. Two of our samples, A/New York/269/2003 (H3N2) and A/New York/32/2003 (H3N2), show that this minor clade continued to circulate in the 2003–2004 season, when most other isolates were reassortants."

According to the CDC

During the 2003–2004 influenza season, influenza A (H1), A (H3N2), and B viruses co-circulated worldwide, and influenza A (H3N2) viruses predominated. Several Asian countries reported widespread outbreaks of avian influenza A (H5N1) among poultry. In Vietnam and Thailand, these outbreaks were associated with severe illnesses and deaths among humans. In the United States, the 2003–2004 influenza season began earlier than most seasons, peaked in December, was moderately severe in terms of its impact on mortality, and was associated predominantly with influenza A (H3N2) viruses.

During September 28, 2003 – May 22, 2004, WHO and NREVSS collaborating laboratories in the United States tested 130,577 respiratory specimens for influenza viruses; 24,649 (18.9%) were positive. Of these, 24,393 (99.0%) were influenza A viruses, and 249

(1.0%) were influenza B viruses. Among the influenza A viruses, 7,191 (29.5%) were subtyped; 7,189 (99.9%) were influenza A (H3N2) viruses, and two (0.1%) were influenza A (H1) viruses. The proportion of specimens testing positive for influenza first increased to >10% during the week ending October 25, 2003 (week 43), peaked at 35.2% during the week ending November 29 (week 48), and declined to <10% during the week ending January 17, 2004 (week 2). The peak percentage of specimens testing positive for influenza during the previous four seasons had ranged from 23% to 31% and peaked during late December to late February.

As of June 15, 2004, CDC had antigenically characterised 1,024 influenza viruses collected by U.S. laboratories since October 1, 2003: 949 influenza A (H3N2) viruses, three influenza A (H1) viruses, one influenza A (H7N2) virus, and 71 influenza B viruses. Of the 949 influenza A (H3N2) isolates characterised, 106 (11.2%) were similar antigenically to the vaccine strain A/Panama/2007/99 (H3N2), and 843 (88.8%) were similar to the drift variant, A/Fujian/411/2002 (H3N2). Of the three A (H1) isolates that were characterised, two were H1N1 viruses, and one was an H1N2 virus. The hemagglutinin proteins of the influenza A (H1) viruses were similar antigenically to the hemagglutinin of the vaccine strain A/New Caledonia/20/99. Of the 71 influenza B isolates that were characterised, 66 (93%) belonged to the B/Yamagata/16/88 lineage and were similar antigenically to B/Sichuan/379/99, and five (7%) belonged to the B/Victoria/2/87 lineage and were similar antigenically to the corresponding vaccine strain B/Hong Kong/330/2001.

H9N2

In December 2003, one confirmed case of avian influenza A (H9N2) virus infection was reported in a child aged five years in Hong Kong. The child had fever, cough, and nasal discharge in late November, was hospitalised for two days, and fully recovered. The source of this child's H9N2 infection is unknown.

H5N1

During January–March 2004, a total of 34 confirmed human cases of avian influenza A (H5N1) virus infection were reported in Vietnam and Thailand. The cases were associated with severe respiratory illness requiring hospitalisation and a case-fatality proportion of 68% (Vietnam: 22 cases, 15 deaths; Thailand: 12 cases, eight deaths). A substantial proportion of the cases were among children and young adults (i.e., persons aged 5–24 years). These cases were associated with widespread outbreaks of highly pathogenic H5N1 influenza among domestic poultry.

H7N3

During March 2004, health authorities in Canada reported two confirmed cases of avian influenza A (H7N3) virus infection in poultry workers who were involved in culling of poultry during outbreaks of highly pathogenic H7N3 on farms in the Fraser River Valley, British Columbia. One patient had unilateral conjunctivitis and nasal discharge, and the other had unilateral conjunctivitis and headache. Both illnesses resolved without hospitalisation.

H7N2

During the 2003–2004 influenza season, a case of avian influenza A (H7N2) virus infection was detected in an adult male from New York, who was hospitalised for upper and lower respiratory tract illness in November 2003. Influenza A (H7N2) virus was isolated from a respiratory specimen from the patient, whose acute symptoms resolved. The source of this person's infection is unknown.

2004 Southern Hemisphere Winter Season

The composition of influenza virus vaccines for use in the 2004 Southern Hemisphere influenza season recommended by the World Health Organisation was:

- an A/New Caledonia/20/99(H1N1)-like virus
- an A/Fujian/411/2002(H3N2)-like virus (A/Kumamoto/102/2002 and A/Wyoming/3/2003 were egg-grown A/Fujian/411/2002-like viruses)
- a B/Hong Kong/330/2001-like virus (B/Shandong/7/97, B/Hong Kong/330/2001 and B/Hong Kong/1434/2002 were among those used at the time. B/Brisbane/32/2002 was also available.)

2004–2005 Northern Hemisphere Winter Season

On the basis of antigenic analyses of recently isolated influenza viruses, epidemiologic data, and postvaccination serologic studies in humans, the Food and Drug Administration's Vaccines and Related Biological Products Advisory Committee (VRBPAC) recommended that the 2004–05 trivalent influenza vaccine for the United States contain A/New Caledonia/20/99-like (H1N1), A/Fujian/411/2002-like (H3N2), and B/Shanghai/361/2002-like viruses. Because of the growth properties of the A/Wyoming/3/2003 and B/Jiangsu/10/2003 viruses, U.S. vaccine manufacturers are using these antigenically equivalent strains in the vaccine as the H3N2 and B components, respectively. The A/New Caledonia/20/99 virus will be retained as the H1N1 component of the vaccine.

2005 Southern Hemisphere Winter Season

The composition of influenza virus vaccines for use in the 2005 Southern Hemisphere influenza season recommended by the World Health Organisation was:

- an A/New Caledonia/20/99(H1N1)-like virus;
- an A/Wellington/1/2004(H3N2)-like virus;
- a B/Shanghai/361/2002-like virus (B/Shanghai/361/2002, B/Jilin/20/2003 and B/Jiangsu/10/2003 were used at the time)

2005–2006 Northern Hemisphere Winter Season

The vaccines produced for the 2005–2006 season use:

- an A/New Caledonia/20/1999-like(H1N1);
- an A/California/7/2004-like(H3N2) (or the antigenically equivalent strain A/New York/55/2004);
- a B/Jiangsu/10/2003-like viruses.

In people in the United States, overall flu and pneumonia deaths were below those of a typical flu season with 84% Influenzavirus A and the rest Influenzavirus B. Of the patients who had Type A viruses, 80% had viruses identical or similar to the A bugs in the vaccine. 70% of the people testing positive for a B virus had Type B Victoria, a version not found in the vaccine.

"During the 2005–06 season, influenza A (H3N2) viruses predominated overall, but late in the season influenza B viruses were more frequently isolated than influenza A viruses. Influenza A (H1N1) viruses circulated at low levels throughout the season. Nationally, activity was low from October through early January, increased during February, and peaked in early March. Peak activity was less intense, but activity remained elevated for a longer period of time this season compared to the previous three seasons. The longer period of elevated activity may be due in part to regional differences in the timing of peak activity and intensity of influenza B activity later in the season."

2006 Southern Hemisphere Winter Season

The composition of influenza virus vaccines for use in the 2006 Southern Hemisphere influenza season recommended by the World Health Organisation was:

- an A/New Caledonia/20/99(H1N1)-like virus;
- an A/California/7/2004(H3N2)-like virus (A/New York/55/2004 was used at the time);
- a B/Malaysia/2506/2004-like virus.

2006–2007 Northern Hemisphere Winter Season

The 2006–2007 influenza vaccine composition recommended by the World Health Organisation on February 15, 2006 and the U.S. FDA's Vaccines and Related Biological Products Advisory Committee (VRBPAC) on February 17, 2006 use:

- an A/New Caledonia/20/99 (H1N1)-like virus;
- an A/Wisconsin/67/2005 (H3N2)-like virus (A/Wisconsin/67/2005 and A/Hiroshima/52/2005 strains);
- a B/Malaysia/2506/2004-like virus from B/Malaysia/2506/2004 and B/Ohio/1/2005 strains which are of B/Victoria/2/87 lineage.

2007 Southern Hemisphere Winter Season

The composition of influenza virus vaccines for use in the 2007 Southern Hemisphere influenza season recommended by the World Health Organisation on September 20, 2006 was:

- an A/New Caledonia/20/99(H1N1)-like virus,
- an A/Wisconsin/67/2005(H3N2)-like virus (A/Wisconsin/67/2005 and A/Hiroshima/52/2005 were used at the time),
- a B/Malaysia/2506/2004-like virus.

2007–2008 Northern Hemisphere Winter Season

The composition of influenza virus vaccines for use in the 2007–2008 Northern Hemisphere influenza season recommended by the World Health Organisation on February 14, 2007 was:

- an A/Solomon Islands/3/2006 (H1N1)-like virus;
- an A/Wisconsin/67/2005 (H3N2)-like virus (A/Wisconsin/67/2005 (H3N2) and A/Hiroshima/52/2005 were used at the time);
- a B/Malaysia/2506/2004-like virus.

In the US, the CDC reported in Feb 2008 that the H1N1 component was a good match (96%) to the infections occurring. But 87% of the H3N2 are A/Brisbane/10/2007-like viruses, which are a recent antigenic variant of the vaccine strain, A/Wisconsin.

And 93% of the B viruses are in a B/Yamagata lineage that is relatively distinct from the vaccine strain B/Victoria lineage. Only one of the three components was a good match; A/Wisconsin is moderately protective against the drifted A/Brisbane strain. 4.5% of those viruses tested are resistant to Oseltamivir, or Tamiflu—a significant increase over previous years.

This vaccine has been described as 40% effective compared to other years that have been 85–95% effective.

2008 Southern Hemisphere Winter Season

The composition of virus vaccines for use in the 2008 Southern Hemisphere influenza season recommended by the World Health Organisation on September 17–19, 2007 was:

- an A/Solomon Islands/3/2006 (H1N1)-like virus;
- an A/Brisbane/10/2007 (H3N2)-like virus;
- a B/Florida/4/2006-like virus

2008-2009 Northern Hemisphere Winter Season

The composition of virus vaccines for use in the 2008-2009 Northern Hemisphere influenza season recommended by the World Health Organisation on February 14, 2008 was:

- an A/Brisbane/59/2007 (H1N1)-like virus;
- an A/Brisbane/10/2007 (H3N2)-like virus;
- a B/Florida/4/2006-like virus (B/Florida/4/2006 and B/Brisbane/3/2007 (a B/Florida/4/2006-like virus) were used at the time).

As of May 30, 2009: "CDC has antigenically characterised 1,567 seasonal human influenza viruses [947 influenza A (H1), 162 influenza A (H3) and 458 influenza B viruses] collected by U.S. laboratories since October 1, 2008, and 84 novel influenza A (H1N1) viruses. All 947 influenza seasonal A (H1) viruses are related to the influenza A (H1N1) component of the 2008-09 influenza vaccine (A/Brisbane/59/2007). All 162 influenza A (H3N2) viruses are related to the A (H3N2) vaccine component (A/Brisbane/10/2007). All 84 novel influenza A (H1N1) viruses are related to the A/California/07/2009 (H1N1) reference virus selected by WHO as a potential candidate for novel influenza A (H1N1) vaccine. Influenza B viruses currently circulating can be divided into two distinct lineages represented by the B/Yamagata/16/88 and B/Victoria/02/87 viruses. Sixty-one influenza B viruses tested belong to the B/Yamagata lineage and are related to the vaccine strain (B/Florida/04/2006). The remaining 397 viruses belong to the B/Victoria lineage and are not related to the vaccine strain."

2009 Southern Hemisphere Winter Season

The composition of virus vaccines for use in the 2009 Southern Hemisphere influenza season recommended by the World Health Organisation on September 17–19, 2008 was:

- an A/Brisbane/59/2007 (H1N1)-like virus;
- an A/Brisbane/10/2007 (H3N2)-like virus;
- a B/Florida/4/2006-like virus.

2009-2010 Northern Hemisphere Winter Season

The composition of virus vaccines for use in the 2009-2010 Northern Hemisphere influenza season recommended by the World Health Organisation on February 12, 2009 was:

- an A/Brisbane/59/2007 (H1N1)-like virus;
- an A/Brisbane/10/2007 (H3N2)-like virus;
- a B/Brisbane/60/2008-like virus.

Since the A/Brisbane/59/2007 (H1N1)-like virus used in the vaccine is an unrelated seasonal strain of influenza, it probably cannot create immunity to the new, non-seasonal strain of influenza A virus subtype H1N1 responsible for the 2009 swine flu outbreak.

2010 Southern Hemisphere Winter Season

- an A/California/7/2009 (H1N1)-like virus;
- an A/Perth/16/2009 (H3N2)-like virus;
- a B/Brisbane/60/2008-like virus.

The H1N1 strain used in this composition is the same strain used in the 2009 flu pandemic vaccine.

2010-2011 Northern Hemisphere Winter Season

The composition of virus vaccines for use in the 2010-2011 Northern Hemisphere influenza season recommended by the World Health Organisation on February 18, 2010 was:

- an A/California/7/2009 (H1N1)-like virus;
- an A/Perth/16/2009 (H3N2)-like virus;
- a B/Brisbane/60/2008-like virus.

The H1N1 strain used in this composition is the same strain used in the 2009 flu pandemic vaccine.

2011 Southern Hemisphere Winter Season

The composition of virus vaccines for use in the 2011 Southern Hemisphere influenza season recommended by the World Health Organisation on September 29, 2010 was:

- an A/California/7/2009 (H1N1)-like virus;
- an A/Perth/16/2009 (H3N2)-like virus;
- a B/Brisbane/60/2008-like virus.

The H1N1 strain used in this composition is the same strain used in the 2009 flu pandemic vaccine.

2011-2012 Northern Hemisphere Winter Season

The composition of virus vaccines for use in the 2011-2012 Northern Hemisphere influenza season recommended by the World Health Organisation on February 17, 2011 was:

- an A/California/7/2009 (H1N1)-like virus;
- an A/Perth/16/2009 (H3N2)-like virus;
- a B/Brisbane/60/2008-like virus.

The H1N1 strain used in this composition is the same strain used in the 2009 flu pandemic vaccine.

Flu Vaccine for Nonhumans

"Vaccination in the veterinary world pursues four goals: (i) protection from clinical disease, (ii) protection from infection with virulent virus, (iii) protection from virus excretion, and (iv) serological differentiation of infected from vaccinated animals (so-called DIVA principle). In the field of influenza vaccination, neither commercially available nor experimentally tested vaccines have been shown so far to fulfil all of these requirements."

Horses

Horses with horse flu can run a fever, have a dry hacking cough, have a runny nose, and become depressed and reluctant to eat or drink for several days but usually recover in two to three weeks. "Vaccination schedules generally require a primary course of 2 doses, 3–6 weeks apart, followed by boosters at 6–12 month intervals. It is generally recognised that in many cases such schedules may not maintain protective levels of antibody and more frequent administration is advised in high-risk situations."

It is a common requirement at shows in the United Kingdom that horses be vaccinated against equine flu and a vaccination card must be produced; the FEI requires vaccination every six months.

Poultry

Poultry vaccines for bird flu are made on the cheap and are not filtered and purified like human vaccines to remove bits of bacteria or other viruses. They usually contain whole virus, not just hemagglutinin as in most human flu vaccines. Purification to standards needed for humans is far more expensive than the original creation of the unpurified vaccine from eggs. There is no market for veterinary vaccines that are that expensive. Another difference between human and poultry vaccines is that poultry vaccines are adjuvated with mineral oil, which induces a strong immune reaction but can cause inflammation and abscesses. "Chicken vaccinators who have accidentally jabbed themselves have developed painful swollen fingers or even lost thumbs, doctors said. Effectiveness may also be limited. Chicken vaccines are often only vaguely similar to circulating flu strains — some contain an H5N2 strain isolated in Mexico years ago. 'With a chicken, if you use a vaccine that's only 85 percent related, you'll get protection,' Dr. Cardona said. 'In humans, you can get a single point mutation, and a vaccine that's 99.99 percent related won't protect you.' And they are weaker than human vaccines. 'Chickens are smaller and you only need to protect them for six weeks, because that's how long they live till you eat them,' said Dr. John J. Treanor, a vaccine expert at the University of Rochester. Human seasonal flu vaccines contain about 45 micrograms of antigen, while an experimental A(H5N1) vaccine contains 180. Chicken vaccines may contain less than 1 microgram. 'You have to be careful about extrapolating data from poultry to humans,' warned Dr. David E. Swayne, director of the agriculture department's Southeast Poultry Research Laboratory. 'Birds are more closely related to dinosaurs.'"

Researchers, led by Nicholas Savill of the University of Edinburgh in Scotland, used mathematical models to simulate the spread of H5N1 and concluded that "at least 95 per cent of birds need to be protected to prevent the virus spreading silently. In practice, it is difficult to protect more than 90 per cent of a flock; protection levels achieved by a vaccine are usually much lower than this." The Food and Agriculture Organisation of the United Nations has issued recommendations on the prevention and control of avian influenza in poultry, including the use of vaccination. A filtered and purified Influenza A vaccine for humans is being developed and many countries have recommended it be stockpiled so if an Avian influenza pandemic starts jumping to humans, the vaccine can quickly be administered to avoid loss of life. Avian influenza is sometimes called avian flu, and commonly bird flu.

Pigs

Swine origin influenza virus (SoIV) vaccines are extensively used in the swine industry in Europe and North America. Most swine flu vaccine manufacturers include an H1N1 and an H3N2 SoIV strains. Swine influenza has become a greater problem in recent decades. Evolution of the virus has resulted in inconsistent responses to traditional vaccines. Standard commercial swine origin flu vaccines are effective in controlling the problem when the virus strains match enough to have significant cross-protection and custom (autogenous) vaccines made from the specific viruses isolated are created and used in the more difficult cases. SoIV vaccine manufacture Novartis paints this picture: "A strain of swine origin influenza virus (SoIV) called H3N2, first identified in the US in 1998, has brought exasperating production losses to swine producers. Abortion storms are a common sign. Sows go off feed for two or three days and run a fever up to 106°F. Mortality in a naïve herd can run as high as 15%."

Dogs

In 2004, Influenza A virus subtype H3N8 was discovered to cause canine influenza. Because of the lack of previous exposure to this virus, dogs have no natural immunity to this virus. However a vaccine is now available.

Computer-assisted Vaccine Design

A new parametre has been defined to quantify the antigenic distance between two H3N2 influenza strains. This parametre was used to measure antigenic distance between circulating H3N2 strains and the closest vaccine component of the influenza vaccine. For the data between 1971 and 2004, the measure of antigenic distance correlated better with efficacy in humans of the H3N2 influenza A annual vaccine than did current measures of antigenic distance such as phylogenetic sequence analysis or ferret antisera inhibition assays. This measure of antigenic distance could be used to guide the design of the annual flu vaccine. The antigenic distance combined with a multiple-strain avian influenza transmission model was used to study the threat of simultaneous introduction of multiple avian influenza strains. Population at Risk (PaR) can be used to quantify the risk of a flu pandemic and to calculate the improvement that a multiple vaccine offers.

Influenza A Virus Subtype H5N1

Influenza A virus subtype H5N1, also known as "bird flu", A(H5N1) or simply H5N1, is a subtype of the influenza A virus which can cause illness in humans and many other animal species. A bird-adapted strain of H5N1, called HPAI A(H5N1) for "highly pathogenic avian influenza virus of type A of subtype H5N1", is the causative agent of H5N1 flu, commonly known as "avian influenza" or "bird flu". It is enzootic in many bird populations, especially in Southeast Asia. One strain of HPAI A(H5N1) is spreading globally after first appearing in Asia. It is epizootic (an epidemic in nonhumans) and panzootic (affecting animals of many species, especially over a wide area), killing tens of millions of birds and spurring the culling of hundreds of millions of others to stem its spread. Most references to "bird flu" and H5N1 in the popular media refer to this strain.

According to the FAO Avian Influenza Disease Emergency Situation Update, H5N1 pathogenicity is continuing to gradually rise in endemic areas but the avian influenza disease situation in farmed birds is being held in check by vaccination. Eleven outbreaks of H5N1 were reported worldwide in June 2008 in five countries (China, Egypt, Indonesia, Pakistan and Vietnam) compared to 65 outbreaks in June 2006 and 55 in June 2007. The "global HPAI situation can be said to have improved markedly in the first half of 2008 but cases of HPAI are still underestimated and underreported in many countries because of limitations in country disease surveillance systems". On December 9, 2010 the WHO announced a total of 510 human cases which resulted in the deaths of 303 people since 2003. A filtered and purified Influenza A vaccine for humans is being developed and many countries have recommended it be stockpiled so, if an Avian influenza pandemic starts jumping to humans, the vaccine can quickly be administered to avoid loss of life. Avian influenza is sometimes called avian flu, and commonly bird flu.

Signs and Symptoms

The avian influenza hemagglutinin binds alpha 2-3 sialic acid receptors, while human influenza hemagglutinins bind alpha 2-6 sialic acid receptors. This means when the H5N1 strain infects humans, it will replicate in the lower respiratory tract, and consequently will cause viral pneumonia. There is as yet no human form of H5N1, so all humans who have caught it so far have caught avian H5N1. In general, humans who catch a humanised influenza A virus (a human flu virus of type A) usually have symptoms that include fever, cough, sore throat, muscle aches, conjunctivitis, and, in severe cases, breathing problems and pneumonia that may be fatal. The severity of the infection depends in large part on the state of the infected person's immune system and whether the victim has been exposed to the strain before (in which case they would be partially immune). No one knows if these or other symptoms will be the symptoms of a humanised H5N1 flu.

The reported mortality rate of highly pathogenic H5N1 avian influenza in a human is high; WHO data indicate 60% of cases classified as H5N1 resulted in death. However, there is some evidence the actual mortality rate of avian flu could be much lower, as there may be many people with milder symptoms who do not seek treatment and are not counted. In one case, a boy with H5N1 experienced diarrhea followed rapidly by a coma without developing respiratory or flu-like symptoms. There have been studies of the levels of cytokines in humans infected by the H5N1 flu virus. Of particular concern is elevated levels of tumour necrosis factor-alpha, a protein associated with tissue destruction at sites of infection and increased production of other cytokines. Flu virus-induced increases in the level of cytokines is also associated with flu symptoms, including fever, chills, vomiting and headache. Tissue damage associated with pathogenic flu virus infection can ultimately result in death. The inflammatory cascade triggered by H5N1 has been called a 'cytokine storm' by some, because of what seems to be a positive feedback process of damage to the body resulting from immune system stimulation. H5N1 induces higher levels of cytokines than the more common flu virus types.

Genetics

The first known strain of HPAI A(H5N1) (called A/chicken/Scotland/59) killed two flocks of chickens in Scotland in 1959; but that strain was very different from the current highly pathogenic strain of H5N1. The dominant strain of HPAI A(H5N1) in 2004 evolved from 1999 to 2002 creating the Z genotype. It has also been called "Asian lineage HPAI A(H5N1)". Asian lineage HPAI A(H5N1) is divided into two antigenic clades. "Clade 1 includes human and bird isolates from Vietnam, Thailand, and Cambodia and bird isolates from Laos and Malaysia. Clade 2 viruses were first identified in bird isolates from China, Indonesia, Japan, and South Korea before spreading westward to the Middle East, Europe, and Africa. The clade 2 viruses have been primarily responsible for human H5N1 infections that have occurred during late 2005 and 2006, according to WHO. Genetic analysis has identified six subclades of clade 2, three of which have a distinct geographic distribution and have been implicated in human infections:

- Subclade 1, Indonesia
- Subclade 2, Europe, Middle East, and Africa (called EMA)
- Subclade 3, China"

A 2007 study focused on the EMA subclade has shed further light on the EMA mutations. "The 36 new isolates reported here greatly expand the amount of whole-genome sequence data available from recent avian influenza (H5N1) isolates. Before our project, GenBank contained only 5 other complete genomes from Europe for the 2004–2006 period, and it contained no whole genomes from the Middle East or northern Africa. Our analysis showed several new findings. First, all European, Middle Eastern, and African samples fall into a clade that is distinct from other contemporary Asian clades, all of which share common ancestry with the original 1997 Hong Kong strain. Phylogenetic trees built on each of the 8 segments show a consistent picture of 3 lineages. Two of the clades contain exclusively Vietnamese isolates; the smaller of these, with 5 isolates, we label V1; the larger clade, with 9 isolates, is V2. The remaining 22 isolates all fall into a third, clearly distinct clade, labelled EMA, which comprises samples from Europe, the Middle East, and Africa. Trees for the other 7 segments display a similar topology, with clades V1, V2, and EMA clearly separated in each case. Analyses of all available complete influenza (H5N1) genomes and of 589 HA sequences placed the EMA clade as distinct from the major clades circulating in People's Republic of China, Indonesia, and Southeast Asia."

Genetic Structure and Related Subtypes

H5N1 is a subtype of the species *Influenza A virus* of the *Influenzavirus A* genus of the *Orthomyxoviridae* family. Like all other influenza A subtypes, the H5N1 subtype is an RNA virus. It has a segmented genome of eight negative sense, single-strands of RNA, abbreviated as PB2, PB1, PA, HA, NP, NA, MP and NS.

HA codes for hemagglutinin, an antigenic glycoprotein found on the surface of the influenza viruses and is responsible for binding the virus to the cell that is being infected. NA codes for neuraminidase, an antigenic glycosylated enzyme found on the surface of the influenza viruses. It facilitates the release of progeny viruses from infected cells. The hemagglutinin (HA) and neuraminidase (NA) RNA strands specify the structure of proteins that are most medically relevant as targets for antiviral drugs and antibodies. HA and NA are also used as the basis for the naming of the different subtypes of influenza A viruses. This is where the *H* and *N* come from in *H5N1*. Influenza A viruses are significant for their potential for disease and death in humans and other animals. Influenza A virus subtypes that have been confirmed in humans, in order of the number of known human pandemic deaths that they have caused, include:

- H1N1, which caused the 1918 flu pandemic ("Spanish flu") and currently is causing seasonal human flu and the 2009 flu pandemic ("swine flu")

- H2N2, which caused "Asian flu"
- H3N2, which caused "Hong Kong flu" and currently causes seasonal human flu

H5N1, ("bird flu"), which is noted for having a strain (Asian-linage HPAI H5N1) that kills over half the humans it infects, infecting and killing species that were never known to suffer from influenza viruses before (e.g. cats), being unable to be stopped by culling all involved poultry - some think due to being endemic in wild birds, and causing billions of dollars to be spent in flu pandemic preparation and preventiveness

- H7N7, which has unusual zoonotic potential and killed one person
- H1N2, which is currently endemic in humans and pigs and causes seasonal human flu
- H9N2, which has infected three people
- H7N2, which has infected two people
- H7N3, which has infected two people
- H10N7, which has infected two people.

Low Pathogenic H5N1

Low pathogenic avian influenza H5N1 (LPAI H5N1) also called "North American" H5N1 commonly occurs in wild birds. In most cases, it causes minor sickness or no noticeable signs of disease in birds. It is not known to affect humans at all. The only concern about it is that it is possible for it to be transmitted to poultry and in poultry mutate into a highly pathogenic strain.

- 1975 – LPAI H5N1 was detected in a wild mallard duck and a wild blue goose in Wisconsin.
- 1981 and 1985 – LPAI H5N1 was detected in ducks by the University of Minnesota conducting a sampling procedure in which sentinel ducks were monitored in cages placed in the wild for a short period of time.
- 1983 – LPAI H5N1 was detected in ring-billed gulls in Pennsylvania.
- 1986 - LPAI H5N1 was detected in a wild mallard duck in Ohio.
- 2005 - LPAI H5N1 was detected in ducks in Manitoba, Canada.
- 2008 - LPAI H5N1 was detected in ducks in New Zealand.
- 2009 - LPAI H5N1 was detected in commercial poultry in British Columbia.

"In the past, there was no requirement for reporting or tracking LPAI H5 or H7 detections in wild birds so states and universities tested wild bird samples independently of USDA. Because of this, the above list of previous detections might not be all inclusive of past LPAI H5N1 detections. However, the World Organisation for Animal Health (OIE) recently changed its requirement of reporting detections of avian influenza. Effective in 2006, all confirmed LPAI H5 and H7 AI subtypes must be reported to the OIE because of their potential to mutate into highly pathogenic strains. Therefore, USDA now tracks these detections in wild birds, backyard flocks, commercial flocks and live bird markets."

High Mutation Rate

Influenza viruses have a relatively high mutation rate that is characteristic of RNA viruses. The segmentation of its genome facilitates genetic recombination by segment reassortment in hosts infected with two different influenza viruses at the same time. A previously uncontagious strain may then be able to pass between humans, one of several possible paths to a pandemic.

The ability of various influenza strains to show species-selectivity is largely due to variation in the hemagglutinin genes. Genetic mutations in the hemagglutinin gene that cause single amino acid substitutions can significantly alter the ability of viral hemagglutinin proteins to bind to receptors on the surface of host cells. Such mutations in avian H5N1 viruses can change virus strains from being inefficient at infecting human cells to being as efficient in causing human infections as more common human influenza virus types. This doesn't mean that one amino acid substitution can cause a pandemic, but it does mean that one amino acid substitution can cause an avian flu virus that is not pathogenic in humans to become pathogenic in humans.

Influenza A virus subtype H3N2 is endemic in pigs in China, and has been detected in pigs in Vietnam, increasing fears of the emergence of new variant strains. The dominant strain of annual flu virus in January 2006 was H3N2, which is now resistant to the standard antiviral drugs amantadine and rimantadine. The possibility of H5N1 and H3N2 exchanging genes through reassortment is a major concern. If a reassortment in H5N1 occurs, it might remain an H5N1 subtype, or it could shift subtypes, as H2N2 did when it evolved into the Hong Kong Flu strain of H3N2.

Both the H2N2 and H3N2 pandemic strains contained avian influenza virus RNA segments. "While the pandemic human influenza viruses of 1957 (H2N2) and 1968 (H3N2) clearly arose through reassortment between human and avian viruses, the influenza virus causing the 'Spanish flu' in 1918 appears to be entirely derived from an avian source".

Prevention

There are several H5N1 vaccines for several of the avian H5N1 varieties, but the continual mutation of H5N1 renders them of limited use to date: while vaccines can sometimes provide cross-protection against related flu strains, the best protection would be from a vaccine specifically produced for any future pandemic flu virus strain. Dr. Daniel Lucey, co-director of the Biohazardous Threats and Emerging Diseases graduate program at Georgetown University has made this point, "There is no H5N1 pandemic so there can be no pandemic vaccine". However, "pre-pandemic vaccines" have been created; are being refined and tested; and do have some promise both in furthering research and preparedness for the next pandemic. Vaccine manufacturing companies are being encouraged to increase capacity so that if a pandemic vaccine is needed, facilities will be available for rapid production of large amounts of a vaccine specific to a new pandemic strain.

Public Health

"The United States is collaborating closely with eight international organisations, including the World Health Organisation (WHO), the Food and Agriculture Organisation of the United Nations (FAO), the World Organisation for Animal Health (OIE), and 88 foreign governments to address the situation through planning, greater monitoring, and full transparency in reporting and investigating avian influenza occurrences. The United States and these international partners have led global efforts to encourage countries to heighten surveillance for outbreaks in poultry and significant numbers of deaths in migratory birds and to rapidly introduce containment measures. The U.S. Agency for International Development (USAID) and the U.S. Department of State, the U.S. Department of Health and Human Services (HHS), and Agriculture (USDA) are coordinating future international response measures on behalf of the White House with departments and agencies across the federal government".

Together steps are being taken to "minimise the risk of further spread in animal populations", "reduce the risk of human infections", and "further support pandemic planning and preparedness". Ongoing detailed mutually coordinated onsite surveillance and analysis of human and animal H5N1 avian flu outbreaks are being conducted and reported by the USGS National Wildlife Health Centre, the Centres for Disease Control and Prevention, the World Health Organisation, the European Commission, and others.

Treatment

There is no highly effective treatment for H5N1 flu, but oseltamivir (commercially marketed by Roche as Tamiflu), can sometimes inhibit the influenza virus from spreading inside the user's body. This drug has become a focus for some governments and organisations trying to prepare for a possible H5N1 pandemic. On April 20, 2006, Roche AG announced that a stockpile of three million treatment courses of Tamiflu are waiting at the disposal of the World Health Organisation to be used in case of a flu pandemic; separately Roche donated two million courses to the WHO for use in developing nations that may be affected by such a pandemic but lack the ability to purchase large quantities of the drug.

However, who Expert Hassan Al-Bushra has said:

> *"Even now, we remain unsure about Tamiflu's real effectiveness. As for a vaccine, work cannot start on it until the emergence of a new virus, and we predict it would take six to nine months to develop it. For the moment, we cannot by any means count on a potential vaccine to prevent the spread of a contagious influenza virus, whose various precedents in the past 90 years have been highly pathogenic".*

Animal and lab studies suggest that Relenza (zanamivir), which is in the same class of drugs as Tamiflu, may also be effective against H5N1. In a study performed on mice in 2000, "zanamivir was shown to be efficacious in treating avian influenza viruses H9N2, H6N1, and H5N1 transmissible to mammals". In addition, mice studies suggest the combination of zanamivir, celecoxib and mesalazine looks promising producing a 50% survival rate compared to no survival in the placebo arm. While no one knows if zanamivir will be useful or not on a yet to exist pandemic strain of H5N1, it might be useful to stockpile zanamivir as well as oseltamivir in the event of an H5N1 influenza pandemic. Neither oseltamivir nor zanamivir can currently be manufactured in quantities that would be meaningful once efficient human transmission starts. In September, 2006, a WHO scientist announced that studies had confirmed cases of H5N1 strains resistant to Tamiflu and Amantadine. Tamiflu-resistant strains have also appeared in the EU, which remain sensitive to Relenza.

Epidemiology

The earliest infections of humans by H5N1 coincided with an epizootic (an epidemic in nonhumans) of H5N1 influenza in Hong Kong's poultry population. This panzootic (a disease affecting animals of many species, especially over a wide area) outbreak was stopped by the killing of the entire domestic poultry population within the territory. However, the disease has continued to spread. On December 21, 2009 the WHO announced a total of 447 cases which resulted in the deaths of 263.

Contagiousness

H5N1 is easily transmissible between birds facilitating a potential global spread of H5N1. While H5N1 undergoes mutation and reassortment, creating variations which can infect species not previously known to carry the virus, not all of these variant forms can infect humans. H5N1 as an avian virus preferentially binds to a type of galactose receptors that populate the avian respiratory tract from the nose to the lungs and are virtually absent in humans, occurring only in and around the alveoli, structures deep in the lungs where oxygen is passed to the blood. Therefore, the virus is not easily expelled by coughing and sneezing, the usual route of transmission.

H5N1 is mainly spread by domestic poultry, both through the movements of infected birds and poultry products and through the use of infected poultry manure as fertilizer or feed. Humans with H5N1 have typically caught it from chickens, which were in turn infected by other poultry or waterfowl. Migrating waterfowl (wild ducks, geese and swans) carry H5N1, often without becoming sick. Many species of birds and mammals can be infected with HPAI A(H5N1), but the role of animals other than poultry and waterfowl as disease-spreading hosts is unknown. According to a report by the World Health Organisation, H5N1 may be spread indirectly. The report stated that the virus may sometimes stick to surfaces or get kicked up in fertilizer dust to infect people.

Virulence

H5N1 has mutated into a variety of strains with differing pathogenic profiles, some pathogenic to one species but not others, some pathogenic to multiple species. Each specific known genetic variation is traceable to a virus isolate of a specific case of infection. Through antigenic drift, H5N1 has mutated into dozens of highly pathogenic varieties divided into genetic clades which are known from specific isolates, but all currently belonging to genotype Z of avian influenza virus H5N1, now the dominant genotype.

H5N1 isolates found in Hong Kong in 1997 and 2001 were not consistently transmitted efficiently among birds and did not cause significant disease in these animals. In 2002 new isolates of H5N1 were appearing within the bird population of Hong Kong. These new isolates caused acute disease, including severe neurological dysfunction and death in ducks. This was the first reported case of lethal influenza virus infection in wild aquatic birds since 1961. Genotype Z emerged in 2002 through reassortment from earlier highly pathogenic genotypes of H5N1 that first infected birds in China in 1996, and first infected humans in Hong Kong in 1997. Genotype Z is endemic in birds in Southeast Asia, has created at least two clades that can infect humans, and is spreading across the globe in bird populations. Mutations are occurring within this genotype that are increasing their pathogenicity. Birds are also able to shed the virus for longer periods of time before their death, increasing the transmissibility of the virus.

Transmission and Host Range

Infected birds transmit H5N1 through their saliva, nasal secretions, feces and blood. Other animals may become infected with the virus through direct contact with these bodily fluids or through contact with surfaces contaminated with them. H5N1 remains infectious after over 30 days at 0 °C (32.0 °F) (over one month at freezing temperature) or 6 days at 37 °C (98.6 °F) (one week at human body temperature) at ordinary temperatures it lasts in the environment for weeks. In Arctic temperatures, it doesn't degrade at all.

Because migratory birds are among the carriers of the highly pathogenic H5N1 virus, it is spreading to all parts of the world. H5N1 is different from all previously known highly pathogenic avian flu viruses in its ability to be spread by animals other than poultry.

In October 2004, researchers discovered that H5N1 is far more dangerous than was previously believed. Waterfowl were revealed to be directly spreading the highly pathogenic strain of H5N1 to chickens, crows, pigeons, and other birds, and the virus was increasing its ability to infect mammals as well. From this point on, avian flu experts increasingly referred to containment as a strategy that can delay, but not ultimately prevent, a future avian flu pandemic.

"Since 1997, studies of influenza A (H5N1) indicate that these viruses continue to evolve, with changes in antigenicity and internal gene constellations; an expanded host range in avian species and the ability to infect felids; enhanced pathogenicity in experimentally infected mice and ferrets, in which they cause systemic infections; and increased environmental stability."

The New York Times, in an article on transmission of H5N1 through smuggled birds, reports Wade Hagemeijer of Wetlands International stating, "We believe it is spread by both bird migration and trade, but that trade, particularly illegal trade, is more important".

On September 27, 2007 researchers reported that the H5N1 bird flu virus can also pass through a pregnant woman's placenta to infect the fetus. They also found evidence of what doctors had long suspected—that the virus not only affects the lungs, but also passes throughout the body into the gastrointestinal tract, the brain, liver, and blood cells.

Current Status of H5N1 Candidate Vaccines

Candidate vaccines were developed in the United States and the United Kingdom during 2003 for protection against the strain that was isolated from humans in Hong Kong in February 2003 but the 2003 strain died out in 2004 making the vaccine of little use. In April 2004, WHO made an H5N1 prototype seed strain available to manufacturers. In August 2006, WHO changed the prototype strains and now offers three new prototype strains which represent three of the six subclades of the clade 2 virus which have been responsible for many of the human cases that have occurred since 2005.

The National Institute of Allergy and Infectious Diseases (NIAID) awarded H5N1 vaccine contracts to Aventis Pasteur (now Sanofi Pasteur) of Swiftwater, Pennsylvania, and to Chiron Corporation of Emeryville, California. Each manufacturer is using established techniques in which the virus is grown in eggs and then inactivated and further purified before being formulated into vaccines.

"A *universal influenza vaccine* could provide protection against all types of influenza and would eliminate the need to develop individual vaccines to specific H and N virus types. Such a vaccine would not need to be reengineered each year and could protect against an emergent pandemic strain. Developing a universal vaccine requires that researchers identify conserved regions of the influenza virus that do not exhibit antigenic variability by strain or over time. A universal vaccine is being developed by the British company Acambis and is being researched by others as well. Acambis announced in early August 2005 that it has had successful results in animal testing. The vaccine focuses on the M2 viral protein, which does not change, rather than the surface hemagglutinin and neuraminidase proteins targeted by traditional flu vaccines. The universal vaccine is made through bacterial fermentation technology, which would greatly speed up the rate of production over that possible with culture in chicken eggs, plus the vaccine could be produced constantly, since its formulation would not change. Still, such a vaccine is years away from full testing, approval, and use." As of July 2007, phase I clinical trials on humans are underway in which a vaccine that focuses on the M2 viral protein "is being administered to a small group of healthy people in order to verify the safety of the product and to provide an initial insight into the vaccine's effect on the human immune system."

In June 2006, the National Institutes of Health (NIH) began enrolling participants in a Phase 1 H5N1 study of an intranasal influenza vaccine candidate based on MedImmune's live, attenuated vaccine technology. Oct 2010 Inovio starts a phase I clinical trial of its H5N1 vaccine (VGX-3400X).

Approved Human H5N1 Vaccines

On April 17, 2007 the US FDA approved "Influenza Virus Vaccine, H5N1" by manufacturer Sanofi Pasteur Inc for manufacture at its Swiftwater, PA facility. In March of 2006, Hungarian Prime Minister Ferenc Gyurcsany reported that Omninvest developed a vaccine to protect humans against the H5N1 influenza strain. The vaccine was approved by the country's national pharmaceutical institute for commercial production.

Results of Trials

Early results from H5N1 clinical trials showed poor immunogenicity compared to the 15 mcg dose that induces immunity in a seasonal flu vaccine. Trials in 2006 and 2007 using two 30 mcg doses produced unacceptable results while a 2006 trial using two doses of 90 mcg each achieved acceptable levels of protection. Current flu vaccine manufacturing plants can not produce enough pandemic flu vaccine at this high dose level.

"Adjuvanted vaccines appear to hold the greatest promise for solving the grave supply-demand imbalance in pandemic influenza vaccine development. They come with obstacles—immunologic, regulatory, and commercial—but they also have generated more excitement than any other type of vaccine thus far. In August 2007, scientists working with a GlaxoSmithKline formula published a trial of a two-dose regimen of an inactivated split-virus vaccine adjuvanted with a proprietary oil-in-water emulsion; after the second injection, even the lowest dose of 3.8 mcg exceeded EU criteria for immune response. And in September, Sanofi Pasteur reported in a press release that an inactivated vaccine adjuvanted with the company's own proprietary formula induced EU-accepted levels of protection after two doses of 1.9 mcg."

The "GlaxoSmithKline-backed team that described an acceptable immune response after two adjuvanted 3.8-microgram (mcg) doses found that three fourths of their subjects were protected not only against the clade 1 Vietnam virus on which the vaccine was based, but against a drifted clade 2 virus from Indonesia as well. To achieve prepandemic vaccines, researchers would have to ascertain the right dose and dose interval, determine how long priming lasts, and solve the puzzle of measuring primed immunity. Further, regulatory authorities would have to determine the trial design that could deliver those answers, the public discussion that would be necessary for prepandemic vaccines to be accepted, and the safety data that would need to be gathered once the vaccines went into use".

Individual Studies

Revaccination - January 2006: The purpose of this study is to determine whether having received an H5 vaccine in the past primes the immune system to respond rapidly to another dose of H5 vaccine. Subjects who participate in this study will have participated in a previous vaccine study (involving the A/Hong/Kong/97 virus) during the fall of 1998 at the University of Rochester.

A/H5N1 in Adult - February 2006

The purpose of this study is to determine the dose-related safety of flu vaccine in healthy adults. To determine the dose-related effectiveness of flu vaccine in healthy adults approximately 1 month following receipt of 2 doses of vaccine. To provide information for the selection of the best dose levels for further studies.

H5 Booster after Two doses - June 2006

The purpose of this study is to determine whether a third dose of vaccines containing A/Vietnam/1203/04 provides more immunity than two doses. Subjects who participate in this study, will have participated in DMID protocol 04-063 involving the A/Vietnam/1203/04. In this study, each subject will be asked to receive a third dose of the H5 vaccine at the same level administered in protocol 04-063.

H5 in the Elderly - August 2006

This study is intended to examine the safety and dose-related immunogenicity of three dosage levels of the Influenza A/H5N1 vaccine, as compared to saline placebo, given intramuscularly to healthy elderly adults approximately 4 weeks apart.

H5 in Healthy Adults - November 2006

This randomised, controlled, double-blinded, dose-ranging, Phase I-II study in 600 healthy adults, 18 to 49 years old, is designed to investigate the safety, reactogenicity, and dose-related immunogenicity of an investigational inactivated influenza A/H5N1 virus vaccine when given alone or combined with aluminum hydroxide. A secondary goal is to guide selection of vaccine dosage levels for expanded Phase II trials based on reactogenicity and immunogenicity profiles. This dose optimisation will be applied to both younger and older subject populations in subsequent studies. Subjects who meet the entry criteria for the study will be enrolled at one of up to 5 study sites and will be randomised into 8 groups to receive two doses of influenza A/H5N1 vaccine containing 3.75, 7.5, 15, or 45 mcg of HA with or without aluminum hydroxide adjuvant by IM injection (N= 60 or 120/vaccine dose group).

Bird flu - November 2006

This study is designed to gather critical information on the safety, tolerability, and the immunogenicity (capability of inducing an immune response) of A/H5N1 virus vaccine in healthy adults. Up to 280 healthy adults, aged 18 to 64, will participate in the study. Each subject will participate for 7 months and will be randomly placed in one of several different study groups receiving a different dose of vaccine, vaccine plus adjuvant, or placebo. All subjects will receive two injections of their assigned study product, about 28 days apart, in their muscle tissue. Subjects will keep a journal of their temperature and any adverse effects between study visits. A small amount of blood will also be drawn before the first injection, 7 days after each injection, and 6 months after the second injection.

Pandemic Flu - January 2007

This Australian study will test the safety and immunogenicity of an H5N1 pandemic influenza vaccine in healthy adults.

Children - February 2007

This is a randomised, double-blinded, placebo-controlled, staged, dose-ranging, Phase I/II study to evaluate the safety, reactogenicity, and immunogenicity of 2 doses of an IM inactivated influenza A/H5N1 vaccine in healthy children, aged 2 through 9 years. This study is designed to investigate the safety, tolerability, and dose-related immunogenicity of an investigational inactivated influenza A/H5N1 vaccine. A secondary goal is to identify an optimal dosage level of the vaccine that generates an acceptable immunogenic response, while maintaining an adequate safety profile.

Bibliography

Aho, M. and J. Hirn. : *Prevalence of Campylobacteria in the Finnish Broiler Chicken Chain from the Producer to the Consumer*, Acta Veterinaira Scandinavica, 1988.

Appleby, M.C & Hughes, B.O. & Elson, H.A.: *Poultry Production Systems, Behaviour, Management and Welfare,* CAB International, NY, 1992.

Berton, V. and Mudd, D.: *Profitable Poultry: Raising Birds on Pasture,* USDA's Sustainable Agriculture Network (SAN), Washington, DC, 2001.

Bourke, M.: *Beak Trimming Handbook for Egg Producers,* Landlinks, UK, 2006.

Chenoweth, H.: *Free-Range Poultry,* Free-Range Poultry Production and Marketing, Creola, Ohio, 2001.

Dennis Wages: *Biosecurity in the Poultry Industry*, International Book, Delhi, 2006.

Elson, H.A.: *Poultry Production Systems, Behaviour, Management and Welfare,* CAB International, NY, 1992.

Fanatico, A.: *Pastured Poultry: A Heifer Project International Case Study Booklet*, Little Rock, AR, 2000.

Frandson, R.D & Wilke, W.L & Fails, A.D.: *Anatomy and Physiology of Farm Animals,* Lippincott, Williams & Wilkins, USA, 2003.

Gillespie, J.R.: *Modern Livestock and Poultry Production 7th edition,* Thomson Delmar Learning, 2004.

Grist, A.: *Poultry Inspection, Anatomy, Physiology and Disease Conditions,* Nottingham University Press, UK, 2006.

Huopalahti, R. & Lopez-Fandino, R. & Anton, M. & Schade, R.: *Bioactive Egg Compounds,* Springer, Delhi, 2007.

Hussain, T.; Jilani, G.; Yaseen, M.; Abbas, M.A.: *Effect of Organic Amendments and EM on Crop Production in Pakistan,* SP, Brazil. Pub. USDA. Washington, D.C., 19991.

Jordan, F & Pattison, M & Alexander, D & Faragher, T.: *Poultry Diseases,* WB Saunders, UK, 2002.

Larbier, M.; Leclercq, B.: *Nutrition and Feeding of Poultry*, Nottingham University Press, U.K., 1994.

Leeson, S & Summers, J.D.: *Scott's Nutrition of the Chicken,* University Books, UK, 2001.

McNab, J.M. & Boorman, K.N.: *Poultry Feedstuffs, Supply Composition and Nutritive Value,* CABI Publishing, UK, 2002.

Mead, G.: *Food Safety Control in the Poultry Industry,* Woodhead Publishing Limited, Abington Hall, Abington, Cambridge, 2006.

Nicholls, C., Paterson, J.: *The Workboot Series: The Story of Chicken,* Kondinin Group, Cloverdale W.A, 2008.

Prasad, L N : *Advanced Pathology and Treatment of Diseases of Poultry : With Special Reference to Etiology Signs*, International Book Dist, Delhi, 2006.

Randall, C.J.: *Color Atlas of Diseases and Disorders of the Domestic Fowl and Turkey,* Iowa State University Press, UK, 1991.

Renema, R.A.: *New Developments in Reproduction and Incubation of Broiler Chickens,* Spotted Cow Press, Ltd, Edmonton, Alberta, Canada, 2003.

Roberts, J.R. & Ball, W.: *Egg Quality Guidelines for the Australian Egg Industry,* AECL Publication, UK, 2004.

Salnsbury, D.: *Poultry Health and Management*, Blackwell Scientific, London, U.K., 1992.

Scanes, C.G. & Brant, G. & Ensminger, M.E.: *Poultry Science,* Pearson Prentice Hill, New Jersey, 1992.

Sharma, R.P. : *Poultry Production in India,* Indian Council of Agricultural Research, Delhi, 2007.

Shukla, Rakesh Kumar : *Handbook of Poultry Diseases : A Bedside-Guide,* International Book Dist, Delhi, 2006.

Stadelman, W.J. & Cotterill, O.J.: *Egg Science and Technology,* Food Products Press, Imprint of Haworth Press, New York, London, 1995.

Thyagarajan, D. : *Diseases of Poultry,* Satish Serial pub, Delhi, 2011.

Tullett, S.G.: *Poultry Science Symposium Number 22.* Butterworth-Heinemann, NJ, 1991.

Verma, S. S.: *Microbiological Changes on Chicken Carcasses During Processing,* Indian Journal of Poultry Science, 1989.

Whitehead, C.C.: *Bone Biology and Skeletal Disorders in Poultry,* Carfax Publishing Company, U.K., 1992.

Index

❑❑❑